1 MONTH OF
FREE
READING

at
www.ForgottenBooks.com

By purchasing this book you are eligible for one month membership to ForgottenBooks.com, giving you unlimited access to our entire collection of over 1,000,000 titles via our web site and mobile apps.

To claim your free month visit:
www.forgottenbooks.com/free912936

ISBN 978-0-266-93975-7
PIBN 10912936

Fifteenth Biennial Report

ISSUED BY THE

BUREAU OF MINES

OF THE

State of Colorado

FOR THE

YEARS 1917 and 1918

FRED CARROLL, Commissioner

DENVER, COLORADO
EAMES BROTHERS, STATE PRINTERS
1919

Fifteenth Biennial Report

ISSUED BY THE

BUREAU OF MINES

OF THE

State of Colorado

FOR THE

YEARS 1917 and 1918
HISTORICAL
COLLECTIONS

FRED CARROLL, Commissioner

DENVER, COLORADO
EAMES BROTHERS, STATE PRINTERS
1919

TABLE OF CONTENTS

LETTER OF TRANSMITTAL
BUREAU OF MINES, STATE OF COLORADO

To His Excellency,
 JULIUS C. GUNTER,
  Governor of Colorado.

Sir: In compliance with the provisions of Section 4268, Revised Statutes of 1908, State of Colorado, I have the honor to submit herewith the Fifteenth Biennial Report of the State Bureau of Mines, which covers the operations of the Bureau for the biennial period 1917-1918.

There is submitted in connection therewith, a statistical report of the accidents causing injuries to persons engaged in the metal mining and allied industries of this state.

And there is also appended a report of the metal production of Colorado by counties, from the beginning of the metal mining industry in 1859, to the close of 1918:

Respectfully,

FRED CARROLL,
 Commissioner of Mines.

January 1, 1919.

PERSONNEL OF THE BUREAU OF MINES OF THE STATE OF COLORADO .

FRED CARROLL, Commissioner..Denver
THOS. R. HENAHEN, Inspector, District No. 1..........................Denver
M. J. McCARTHY, Inspector, District No. 2.................Cripple Creek
R. J. MURRAY, Inspector, District No. 3..................................Leadville
ROBERT INNES, Inspector, District No. 4..................................Durango
J. T. DUCE, Chief Clerk..Boulder
MRS. A. M. NICKERSON, Stenographer........................... ....Denver

Mr. Henahen succeeeded Mr. Thos. Dunstone, of Blackhawk, as Inspector for District No. 1, September 1st, 1917.

Mr. Murray succeeded Mr. A. E. Moynahan, of Alma, as Inspector for District No. 3, September 1st, 1917.

Mr. Duce succeeded Mr. Tomblin, of Denver, as Chief Clerk, January 1st, 1918.

DISTRICTS OF INSPECTION

DISTRICT No. 1—Denver, Jefferson, Boulder, Larimer, Jackson, Routt, Grand, Gilpin, Clear Creek and Moffat counties.

DISTRICT No. 2—El Paso, Teller, Pueblo, Las Animas, Huerfano, Custer and Fremont counties.

DISTRICT No. 3—Lake, Summit, Chaffee, Park, Pitkin, Mesa, Delta, Eagle, Rio Blanco and Garfield counties.

DISTRICT No. 4—San Juan, Ouray, Hinsdale, Mineral, Rio Grande, Saguache, Costilla, Conejos, Archuleta, La Plata, Montezuma, Dolores, San Miguel, Montrose and Gunnison counties.

FINANCIAL STATEMENT OF BUREAU OF MINES
1917-1918

RECEIPTS.

Appropriation for the fiscal years 1917-1918:

Salary, Commissioner	$ 6,000.00	
Salaries, Inspectors	14,400.00	
Salary, Clerk	3,000.00	
Salary, Stenographer	2,400.00	
		$25,800.00
Traveling expenses, Commissioner	$ 2,000 00	
Transfer to traveling expense, Inspectors	276.28	
		$ 1,723.72
Traveling expenses, Inspectors	$ 6,000.00	
Transferred from Commissioner's Trav. Exp.	276.28	
		$ 6,276.28
Incidental Fund	$ 1,000.00	1,000.00
		$34,800.00

DISBURSEMENTS.

	Salary	Expense	Incidental	
Commissioner	$ 6,000.00	$1,714.54		
Inspector Dist. No. 1	3,600.00	1,559.30		
Inspector Dist. No. 2	3,600.00	1,520.08		
Inspector Dist. No. 3	3,600.00	1,298.21		
Inspector Dist. No. 4	3,600.00	1,835.84		
Chief Clerk	3,000.00			
Stenographer	2,400.00			
Incidental			$ 926.59	
	$25,800.00	$7,927.97	$ 926.59	$34,654.56
Unexpended balance		72.03	73.41	145.44
				$34,800.00

INTRODUCTION

The past two years have been unusually exacting ones for the mining industry. The successful prosecution of the war depended much upon the maintenance of a supply of metals. But early in 1917 the Government adopted a definite policy of price-fixing, which was applied most rigorously to mineral products. Thus the mine operator was caught between the Scylla and Charybdis of increasing cost of supplies and the decreasing efficiency of labor and a fixed price for his output. It is much to the credit of the mining industry that under such adverse conditions the production was maintained.

This report contains tables and graphs which show at a glance the present condition of the industry. Production for 1918 was influenced to some extent also by the epidemic of Spanish Influenza, which closed many mines in October and November, and by the scarcity of labor.

ACKNOWLEDGMENTS

An expression of special appreciation is due to George Otis Smith, Washington, D. C., and C. W. Henderson, Denver, Colorado, Director and Statistician, respectively, of the United States Geological Survey, for the county production statistics published herein. This is the first publication of this very important information and it was only through the courtesy of Mr. Smith and the help of Mr. Henderson that we were able to get the use of data which it took Mr. Henderson several years to collect and compile.

We are also indebted to Director Van H. Manning and Statistician A. H. Fay of the Federal Bureau of Mines for valuable suggestions and standardized forms used in the preparation of accident statistics, to the State Industrial Commission for valuable assistance, and access to their accident records, and to the mine operators of this state for labor data. In addition, I also wish to express my appreciation of the faithful and efficient service rendered by those associated with me in this department.

THE MINERAL COLLECTION

About one thousand specimens were added to the collection during the past biennial period. Of special interest are some very rich gold and silver ores contributed by John G. Morgan, Mrs. Allen T. Wells and Costin Brothers and Casin, and also a number of other specimens obtained by exchange from New York and New Jersey. The collection of the Colorado Scientific Society, which was in very bad shape, was catalogued. It seemed important to preserve this collection from destruction, as it contains a number of specimens from which some rare Colorado minerals were described. This collection is at present housed in the rooms of the Bureau, but we have no agreement for its permanent incorporation in our collection. No special effort was made to increase the size of the collection, as the facilities for taking care of the minerals are lacking.

From time to time the museum has been called upon to supply specimens from inaccessible mines, and also minerals for comparison. Prospectors have brought in much material to compare with specimens of rare minerals which they have never seen. It is expected that this type of work will greatly increase. It is estimated that 80,000 people visited the museum during the past two years.

The cases in which the minerals are displayed are old-fashioned and not dust-proof. They should be replaced as soon as possible, for specimens in them soon become dusty and do not show to good advantage. Furthermore, the collection has already outgrown its present quarters and if the plans developed for its enlargement are carried through, the space given to it must be increased. The present systematic collection of minerals lacks fifty per cent of the known species of minerals. These should be secured. It is also proposed to assemble suites of minerals and rocks to represent the ore occurrences and geology of each district in the state and to collect maps of the mines and geology of these districts besides making similar collections from the most important mining districts of the world. Collections to illustrate the physical properties of minerals are also badly needed. The collection of rocks is far too small and needs much systematic work. To do this work, one man skilled in mineralogy and petrography should be employed solely in the museum and he should be provided with money for adequate laboratory facilities, traveling expenses when in the field, and for the purchase of specimens.

ACCIDENTS

The mining accidents for the years 1917-1918 are tabulated with the previous year's totals for comparison.

Table No. 1 classifies, according to cause, all accidents in the metal mining and quarrying industry. It will be noted that during the year 1917, 56 persons were killed, and 713 seriously injured, as compared with 69 persons killed, and 742 seriously injured during the previous year, while it will be noted further that in 1918, 47 persons were killed and 628 seriously injured.

Table No. 2 shows classification according to cause and occupation by groups of the fatal accidents in the metal mining and quarrying industry of Colorado, during the years 1917 and 1918, and for the years 1915 and 1916, for comparison.

Table No. 3 gives the total days of employment in metal mines and allied industries, and the number of men killed and injured per 10,000 days of employment, during the years 1915, 1916, 1917 and 1918.

CLASSIFICATION ACCORDING TO CAUSE OF ALL ACCIDENTS IN THE METAL MINING AND QUARRYING INDUSTRY IN THE STATE OF COLORADO DURING THE YEARS 1917 AND 1918

UNDERGROUND	Fatal		Serious Injuries—(Time lost over 14 days)						Slight Injuries (Time lost, 1 to 14 days)	
			Permanent Total Disability		Permanent Partial Disability		Temporary Disability			
	1917	1918	1917	1918	1917	1918	1917	1918	1917	1918
1. Falls of rock or ore from roof or wall	13	17	----	----	2	1	81	84	101	66
2. Handling rock or ore	1								5	5
(a) Loading at face	1				1	2	6	9		
(b) Loading at chute							22	16	39	21
(c) Sledging									1	
3. Timber or hand tols	1	1			1		14	14	27	30
4. Explosives—										
(a) Transportation										
(b) Charging	2	1	2		1	1	2	1		1
(c) Suffocation		2						3	5	
(d) Drilling into old holes										
(e) Striking in loose rok				1						
(f) Thawing				1	1		1	1		
(g) Caps, detonators, etc.	1				1		2			
(h) Unguarded shots										
(i) Returned too soon						1	2	2	1	1
(j) Premature shot							4	3	1	1
(k) Miscellaneous							1	1	1	
5. Haulage—										
(a) Hand and animal	1				7	1	48	52	73	58
(b) Mal	1				1		16	7	5	
6. Persons falling down chute, winze, raise, or stope	6	4					39	44	50	33
7. Run of ore from chute or pocket		2					2	1	3	1
8. Drilling (by machine or hand drills)					2	2	20	15	33	18
9. Electricity—										
(a) Direct contact with trolley wire										
(b) Tol or bar striking trolley wire						1			2	1
(c) Contact with motor					1	2			2	
(d) ors	1						1	1	5	3
10. Machinery ther than 5 and 8							1	4		
11. Mine fires							6			
12. Suffocation from natural gases										
13. ash of water	1									
14. Nails and splinters							2	4	16	7
15. ther causes—										
(a) Falling objects, other than 1 and 2	1				4		23	12	50	26
(b) Flying objects, other than 2c					2	1	16	20	38	33
(c) Burns		2					2	1	11	9

	1	2	3	4	5	6	7	8	9	10
(d) Miscellaneous	16	21	8	12	2	1			2	
Total	328	490	303	319	14	24	2	2	31	28
SHAFT										
16. Falling down shaft	3	7	2	3	1	1			1	3
17. Objects falling down shaft				2					1	4
18. Breaking of cables										2
19. Overwinding				1						
20. Cage, skip or bucket—										
(a) Runaway										
(b) Riding with rock or ore										
(c) Riding with timber or tools	1	16	3						1	
(d) Struck by	6	4	7	10	7					
21. Other causes			3						1	5
Total	10	27	15	16		1			3	14
SURFACE SHOPS AND PLANT										
22. Haulage—										
(a) Hand and animal	4	12	3	4						
(b) Coal	5	5	7	8						
23. Railway cars and cages				2		1				1
24. Run or fall of ore in or from ore bins		1		1		1				1
25. Falls of persons	17	17	17	17						
26. Nails and splinters	7	5	1							
27. Hand tools, saws, axes, etc.	7	5	3	4						
28. Electricity—										
(a) Direct contact with trolley wire										
(b) Tool or bar striking trolley wire										
(c) Contact with motor		2								
(d) Others	2	2	2	1					2	2
29. Machinery	4	15	7	16	5				2	4
30. Other causes—										
(a) Falling objects	9	11	9	6						
(b) Flying objects	19	10	4	3						
(c) Burns	2	6	3	2						
(d) Miscellaneous	12	11	8	9						
Total	88	102	64	73	9	7			5	
PLACER MINES—DREDGING										
1. Machinery	1	3		1	2					
2. Electricity	1	1	1							
3. Boiler explosions or bursting of steam pipes										
4. Falls of persons	1	1	1							
5. Hand tools	2	2								1
6. Other causes	5	4	2	1	2					
Total	10	11	4	2						1

CLASSIFICATION ACCORDING TO CAUSE OF ALL ACCIDENTS IN THE METAL MINING AND QUARRYING INDUSTRY IN THE STATE OF COLORADO DURING THE YEARS 1917 AND 1918—Continued.

	Fatal		Serious Injuries—(Time lost over 14 days)						Slight Injuries (Time lost, 1 to 14 days)	
			Permanent Total Disability		Permanent Partial Disability		Temporary Disability			
ORE-DRESSING AND MILLING	1917	1918	1917	1918	1917	1918	1917	1918	1917	1918
1. Haulage system— (a) Cars and motors							2	8	7	
(b) Mechanical conveyors								2		1
2. Railway cars and locomotives	1						1			1
3. Crushers, rolls, or stamps	2				1	1	3	2	7	1
4. Tables, jigs, etc.							2	2		
5. Other machinery		1							14	4
6. Falls of persons					3	8	14	13	21	16
7. Suffocation in ore bins	2				1		28	17		1
8. Falling objects (rocks, timbers, etc.)							12	8	23	14
9. Cyanide or other poisoning								1		1
10. Scalding (steam or water)									1	
11. Electricity	2	1			1		1	1	5	3
12. Hand tools, bars, etc.							1	3	9	17
13. Nails, splinters, etc.								2	10	9
14. Flying pieces of rock from sledging or crushing		1			1		3	1	10	11
15. Other causes					2		9	7	22	17
Total	7	2			9	9	76	67	129	94
SMELTER										
16. Haulage system— (a) Cars and motors		1			2	2	23	22	55	32
(b) Mechanical conveyors		2								
17. Railway cars and locomotives					1	1	6	8	6	10
18. Cranes									3	
19. Other machinery	1	1				1		1		1
20. Falls of persons					1		2	8	4	3
21. Suffocation in ore bins					1	1	8	11	13	19
22. Flying or falling objects (rocks, timbers, etc.)										
23. Gas or asphyxiation					3	1	31	18	59	38
24. Scalding (steam or water)										2
25. Electricity									2	1
26. Hand tools, bars, etc.		1					12	10	11	1
27. Nails, splinters, etc.								1	3	15
28. Burns from matte, slag, or molten metal (pouring or spilling)							25	14	29	18

7	3	8	2					
7	22	4	4					1
154	210	105	113	7	8		5	1

29. Hot-metal explosions
30. Other causes

Total

AUXILIARY WORKS

31. Haulage systems, cars, ..., etc.
32. Railway cars and ...
33. Falls of persons
34. Falling objects (rocks, timbers, etc.)
35. Nails, splinters, etc.
36. Hand tools, ..., bars, etc
37. Electricity
38. Machinery
39. Failure of ladder, scaffold, or other support
40. Handling hot materials
41. Other causes

Total

IN AND ABOUT QUARRY

1. Falls or slides of rock or overburden
2. Handling rock at face
3. ... or hand tools
4. Explosives—
   (a) Transportation
   (b) Charging
   (c) Drilling into old holes
   (d) Striking in loose rock
   (e) ...
   (f) Caps, detonators, etc.
   (g) Unguarded shots
   (h) Returned too soon
   (i) Premature shot
   (j) Miscellaneous
5. Haulage—
   (a) Hand and animal
   (b) ...
6. Falling into ... from surface benches, or face
7. Falling from hoists, derricks, ..., etc.
8. Drilling and channeling (by machine or hand)
9. Electricity (shock or burns)
10. Machinery—
    (a) ... and attachments
    (b) ..., masts, ..., and attachments
    (c) Pumps and hoisting engines

CLASSIFICATION ACCORDING TO CAUSE OF ALL ACCIDENTS IN THE METAL MINING AND QUARRYING INDUSTRY IN THE STATE OF COLORADO DURING THE YEARS 1917 AND 1918—Concluded

| | Fatal | | Serious Injuries—(Time lost over 14 days) | | | | | | Slight Injuries (Time lost, 1 to 14 days) | |
| | | | Permanent Total Disability | | Permanent Partial Disability | | Temporary Disability | | | |
IN AND ABOUT QUARRY—Continued	1917	1918	1917	1918	1917	1918	1917	1918	1917	1918
11. Flying pieces of rock from sledging									4	4
12. Nails, splinters, etc.										
13. Boiler and air-tank explosions										
14. Animals								1		1
15. Others causes							11		9	1
Total						1	43	15	42	18
OUTSIDE WORKS										
16. Haulage—										
(a) Hand and animal	1							5	1	1
(b) Rail motives										
17. Railway cars and motives										
18.										
19. ...ts, derricks, etc.										1
20. ...r machinery								3		1
21. Falls of persons								1		
22. Hand ...ls								1		1
23. Electricity (shock or burns)										
24. Nails, splinters, etc.										
25. Flying pieces of rock from sledging or crushing										
26. Flying or falling objets (rocks, timbers, etc.)										1
27. Burns										
28. ...r causes										
Total	1							10		5

CLASSIFICATION ACCORDING TO CAUSE AND OCCUPATION OF THE FATAL ACCIDENTS IN THE MINING INDUSTRY FOR THE YEARS 1915, 1916, 1917, AND 1918.

MINES.

Date of Accident—	Superintendents and Foremen				Machine Men and Helpers				Miners				Loaders, Shovelers, Muckers and Trammers				Trackmen, Pipemen, Pumpmen and Compressormen				Cage Tenders and Station Cagers				Carpenters, Timbermen and Helpers				Electricians, Motormen and Helpers			
	1915	1916	1917	1918	1915	1916	1917	1918	1915	1916	1917	1918	1915	1916	1917	1918	1915	1916	1917	1918	1915	1916	1917	1918	1915	1916	1917	1918	1915	1916	1917	1918
UNDERGROUND																																
1. Falls of rock or ore from roof or wall	1	3	1		1	4	3	2	5	5	3		2	5	5	11									4	2	7	4				
2. Falls of rock or ore														1		1																
3. Bar or hand tools	2	1			4	4		1		1	3	1	1	2	1	2										1			1	1	1	
4. Explosives													2	3	2	2																
5. Haulage																																
6. Persons falling down mine, winze, raise, or slope	1				3		1	3	3	1	1		3		3	1									1	3	1	1				
7. Run of ore from chute or pocket								1					1			1										2						
8. Drilling (by machine or hand drills)																																
9. Electricity		1			1																											
10. Machinery other than 5 & 8					1																											
11. Mine fires																																
12. Suffocation from natural gases																																
13. Inrush of water																																
14. Nails and splinters												1		2		1																
15. Other causes							1									3	1		1													
Total	4	5	1		10	8	5	7	8	7	7	4	9	12	5	15	1		1	1					5	8	8	5	1	1	1	
SHAFTS																																
16. Falling down shaft	1	1			1	1			1	1	1	1	2	2	1	1						3	1	1	2	1	1	1				
17. Objects falling down shaft														2	1								1	1	1		1					
18. Breaking of cables										2			2	2	1	1						2	2									
19. Overwinding																												1				
20. Cage, skip or bucket		1								3	3																1	1				
21. Other causes	1																	1			1	1										
Total	1	2			1	1			1	5	5	1	2	4	2	1	1	1	1		3	3	3		2	1	3	2				

CLASSIFICATION ACCORDING TO CAUSE AND OCCUPATION OF THE FATAL ACCIDENTS IN THE METAL MINING AND QUARRYING INDUSTRY FOR THE YEARS 1915, 1916, 1917 AND 1918—Continued

SURFACE ACCIDENTS

Date of Accident—	Superintendents and Foremen				Engineers, Firemen and Hoistmen				Blacksmiths and Mechanics				Carpenters and Topmen				Electricians				Others			
	1915	1916	1917	1918	1915	1916	1917	1918	1915	1916	1917	1918	1915	1916	1917	1918	1915	1916	1917	1918	1915	1916	1917	1918
22. Haulage	1	1																						
23. Railway cars and locomotives	1	1																						
24. Run or fall of ore in or from ore bins																								
25. Falls of persons				1																				
26. Nails and splinters													1											
27. Hand tools, bars, etc.													1	2		1	2					1		1
28. Electricity		1											1				1			1				
29. Machinery						1																		
30. Other causes										2				2		2							2	
Total	2	1		1		1				2			2	4		3	3			1		1	3	1

CLASSIFICATION ACCORDING TO CAUSE AND OCCUPATION OF THE FATAL ACCIDENTS THAT OCCURRED IN THE METAL MINING AND QUARRYING INDUSTRY OF COLORADO IN THE YEARS 1915, 1916, 1917, AND 1918—Continued

MILLS

Date of Accident—	Superintendents and Foremen 1915	1916	1917	1918	Grinding Machinery Men 1915	1916	1917	1918	Concentrator Men 1915	1916	1917	1918	Laborers 1915	1916	1917	1918	Carpenters and Mechanics 1915	1916	1917	1918	Aerial Tram Men and Conveyor Men 1915	1916	1917	1918
1. Haulage system		1											1											
2. Railway cars and locomotives																							1	
3. Crushers, rolls, or stamps					2		1																1	
4. Mills, jigs, etc.					1		1					1												
5. Other machinery																								
6. Falls of persons										2														
7. Suffocation in ore bins																								
8. Falling objects (rocks, timbers, etc.)		1																						
9. Cyanide or other poisoning																								
10. Scalding (steam or water)																								
11. Electricity											1								1	1				
12. Hand tools, bars, etc.																								
13. Nails, splinters, etc.																								
14. Flying pieces of rock from sledging or crushing	1												1						1	1				
15. Other causes												1												
Totals	1	2			3		2			2	2	1	1						1	1			2	

CLASSIFICATION ACCORDING TO CAUSE AND OCCUPATION OF THE FATAL ACCIDENTS THAT OCCURRED IN THE METAL MINING AND QUARRYING INDUSTRY OF COLORADO IN THE YEARS 1915, 1916, 1917 AND 1918—Continued.

SMELTERS

Date of Accident—	Superintendents and Foremen				Laborers				Motormen				Feeders and Furnace Men and Charge Wheelers			
	1915	1916	1917	1918	1915	1916	1917	1918	1915	1916	1917	1918	1915	1916	1917	1918
1. Haulage system												1				1
2. Railway cars and locomotives																
3. Cranes																
4. Other machinery																
5. Falls of persons																
6. Suffocation in ore bins																
7. Flying or falling objects (rocks, timbers, etc.)						1	1	1								
8. Gas (fumes or asphyxiation)																
9. Scalding (steam or tar)																
10. Electricity								1								
11. Hand tools, axes, bars, etc.																
12. Nails, splinters, etc.																
13. Burns from fire, slag, or molten metal (pouring or spilling)													2			
14. Hot-metal explosions	*1				*2											
15. Other causes																
Totals	1				2	1	1	3				1	2			1

*Killed by hot tar on account of a defective hose connection.

CLASSIFICATION ACCORDING TO CAUSE AND OCCUPA-
TION OF THE FATAL ACCIDENTS THAT OCCURRED
IN THE METAL MINING AND QUARRYING INDUSTRY
OF COLORADO FOR THE YEARS 1915, 1916, 1917 AND
1918.—Concluded.

QUARRIES

	Teamsters				Drillmen				Laborers			
	1915	1916	1917	1918	1915	1916	1917	1918	1915	1916	1917	1918
Haulage	....	....	1	....	....	....	....	....	....	....	....	....
Machinery	....	....	....	....	....	1	....	....	....	....	....	....
Other causes	....	....	....	....	....	....	....	....	....	1	....	1
Total	....	....	1	....	....	1	....	....	....	1	....	1

PLACER MINES

	Deckhands				Laborers			
	1915	1916	1917	1918	1915	1916	1917	1918
Machinery	....	....	....	....	1	....	....	....
Falls of Persons	....	....	1	....	....	....	....	....
Total	....	....	1	....	1	....	....	....

DAYS OF EMPLOYMENT IN AND ABOUT THE MINES, MILLS AND SMELTERS, AND NUMBER OF MEN KILLED AND INJURED PER 10,000 DAYS OF EMPLOYMENT DURING THE YEARS 1917 AND 1918.

| | Days of Employment | | Killed 1917 | | Killed 1918 | | Serious Injury Time Lost Over 14 Days 1917 | | Serious Injury Time Lost Over 14 Days 1918 | | Slight Injury Time Lost Under 14 Days 1917 | | Slight Injury Time Lost Under 14 Days 1918 | |
	1917	1918	No. of Accidents	Rate Per 10,000 Days	No. of Accidents	Rate Per 10,000 Days	No. of Accidents	Rate Per 10,000 Days	No. of Accidents	Rate Per 10,000 Days	No. of Accidents	Rate Per 10,000 Days	No. of Accidents	Rate Per 10,000 Days
Underground	2,086,405	2,084,696	44	.09	34	.11	68	1.954	36	1.779	51	2.736	66	1.798
Surface	459,308	442,239	4	.90	6	.96	80	1.740	83	1.876	02	2.416	93	2.102
Mills	513,701	565,648	7	.1362	2	.0353	85	1.654	76	1.343	129	2.511	94	1.661
Smelters	853,361	801,270	1	.0411	5	.0624	130	1.523	113	1.410	219	2.566	156	1.946
			56		47		703		628		1021		709	

The Bureau of Mines has taken particular pains to keep its records of accidents up to date, and while a few of the minor accidents may not have been reported, all of the fatal accidents have certainly been reported.

A glance at the following tables will show that the number of accidents reported to this department was less than that of the previous year.

During 1918 there occurred a total of 1,384 accidents, 47 of which were fatal, 628 serious (time lost over 14 days), and 709 slight (time lost 1 to 14 days), with a total of 35,708 days lost. During the previous year there occurred a total of 1,780 accidents, 56 of which were fatal, 703 serious (time lost over 14 days) and 1,021 slight (time lost 1 to 14 days), with a total of 37,436 days lost.

The fatal accidents were divided as follows: Underground 31, shaft 3, surface 5, placers none, quarries 1, mills 2 and smelters 5, in 1918, while the previous year shows, underground 28, shaft 4, surface 4, placers 1, quarries 1, mills 7, smelters 1. This is a good record when the great number of inexperienced men that were employed in the industry is taken into consideration.

Reference to the following tables will show that falls of ground and falls of persons are the chief hazards in mining.

Out of a total of 31 men killed underground during 1918, 17 were killed by falls of ground and 4 by falling down chutes, winzes and raises; and of 3 shaft accidents, one man was killed by falling down a shaft. During the previous year 28 men were killed underground, of which 13 were killed by falls of ground and 7 by falling down chutes, winzes and raises and of 14 shaft accidents 3 men were killed by falling down the shaft. Carelessness in barring down loose ground and the absence of proper guards around shafts, chutes, winzes and raises are the principal causes of these accidents.

Haulage accidents occur frequently and often result in death. Chutes placed on the wrong side of drifts and trammers following too closely and running into each other are the principal causes.

At mills, placers and quarries most accidents are caused by falls of persons and falling objects.

The haulage system and falls of persons are the causes of the most serious accidents at the smelters.

The decrease in the number of slight accidents is due to the growing disposition of employers to take advantage of the ten-day period allowed by law for the reporting of accidents, and if the slightly injured man makes a complete recovery in that time the accident is not reported to the Industrial Commission.

Notes Upon the Various Fatal Accidents During 1917-1918

UNDERGROUND—1917

January 13, at Telluride, while employed as a pipeman, William Matthews, age 50, attempted to jump across the top of an open winze, but lost his footing, fell through the winze 150 feet, and sustained fatal injuries. The mine where this accident occurred is worked through a cross-cut tunnel, and by open and shrinkage stopes and timbered with stulls.

January 14, at Breckenridge, while shoveling dirt from a drift that had just been started in the hanging wall of an inclined shaft, Adolph Winslow, age 28, a machine drill runner, was caught by a run of ground which came through an opening in the cribbing, and crushed to death. The mine where this accident occurred is operated through a cross-cut tunnel and an incline shaft, and by shrinkage and waste filled stopes and is timbered with stulls and cribbing.

January 26, near Cowdry, Jackson County, while engaged as a miner, E. G. Delaney, age 26, was spitting a round of holes in a drift and remained until the first explosion occurred. He was struck by flying rocks that fractured his skull and caused his death within a few hours. The mine where this accident occurred is operated through a shaft 170 feet deep.

February 5, at Nederland, Martin Krogh, age 49, a foreman miner, while gadding ore from a large rock, was caught by a cave from the hanging wall, and pinned against the foot-wall. The internal injuries he received caused his death four hours later. The mine where this accident occurred was a small lease operated through a shaft by waste filled stopes and is timbered with stulls.

February 8, at Telluride, while engaged as machine drill runner, John Morsen, age 35, was caught and instantly killed by a rock that fell from the back of a stope. The mine where this accident occurred is operated through tunnels by shrinkage stopes, and is timbered with stulls.

February 26, in the Cripple Creek District, while attempting to get a ladder out of an old stope, Charles Fightmaster, a machine drill runner, stepped on a sprag that broke under his weight, and let him fall 80 feet into an open stope, causing his death. The mine where this accident occurred is operated through shafts and by shrinkage stopes and is timbered with stulls.

March 7, at Creede, while engaged with another man as a hand miner, Phil. Ragan, age 48, was killed in an upraise by remaining too long at a blast.

March 11, at Telluride, while engaged as a trammer, George Posis, age 33, fell down an open ore chute a distance of 60 feet and suffered injuries that proved fatal. The mine where this accident occurred is operated through tunnels, upraises and by shrinkage stopes and is timbered with stulls.

April 6, at Red Cliff, while engaged as a timberman's helper, M. O. Spurr, age 26, was buried by a fall of ore which suffocated him. The mine where this accident occurred was operated by cross-cut tunnels and by breast stopes and is timbered with square sets.

April 6, at Red Cliff, while engaged as a timberman, M. C. Walter, age 23, was buried by a fall of ore and suffocated. The mine where this accident ocurred was operated through cross-cut tunnel and by breast stopes and is timbered with square sets.

May 4, in the Cripple Creek District, while timbering over an open stope, Daniel Connelley, age 54, was thrown 45 feet into the stope by a falling rock; his injuries proved fatal. The mine where this accident occurred is operated through shafts and by open and shrinkage stopes and is timbered with stulls.

May 14, in the Cripple Creek District, while engaged as a mine leaser, Henry E. Hoffman, age 56, was walking along a drift over an open stope that was supported by stulls, when some of the timbers gave way, and allowed him to fall 85 feet with fatal results. The mine where this accident occurred is operated through shafts and by shrinkage and open stopes and is timbered with stulls.

May 17, at Aspen, while engaged as a timberman, Frank Godec, age 38, was caught by a slab of rock which caved from the back of a drift, and instantly killed. The mine where this accident occurred is operated through shafts and by breast stopes and is timbered with square sets.

May 18, at Telluride, while engaged as a timberman's helper, August Gartner, age 35, was caught by a fall of rock from the back of a drift and instantly killed. The mine where this accident occurred is operated through tunnels, and by waste filled stopes and is timbered with stulls.

June 9, at Silverton, while engaged in barring down loose rock, August Beamer, age 31, a foreman miner, was crushed to death by a fall of rock from the back of a stope. The mine where this accident occurred is operated through a shaft and by overhead open stopes and is timbered with stulls.

June 16, at Leadville, while engaged as a trammer, Edward Curran, age 45, was caught in a drift by a car coming from the rear, and fatally injured. The mine where this accident occurred is operated through shafts, and by breast stopes and is timbered with cribs and square sets.

June 20, at Georgetown, while engaged as a timberman, Ferdinand C. Bellotti, age 28, was killed by a slab of rock that fell from the back of a drift. The mine where this accident occurred is operated through a cross-cut tunnel and by waste filled stopes and timbered with cribs and tunnel sets.

July 25, at Naturita, while engaged as a miner, J. E. Myers, age 60, was injured by a fall of ground from the back of a drift, and later died from his injuries The mine where this accident occurred is operated through drifts and by breast stopes and timbered with props

August 15. at Telluride, in order to escape from a winze where he was engaged as a machine drill runner and avoid the danger of a rush of water from old workings that had been tapped by a raise, Frank Koivu, age 35, started to climb out of the winze and was either electrocuted or overcome with water and gas and died standing on the ladder.

September 11, at Telluride, while riding in an empty car drawn by a motor, Maguel Salzar, age 19, a car loader, was caught between a car and chute and his skull fractured so that he died later. The mine where this accident occurred is operated through a tunnel, by waste filled stopes, and is timbered with stulls and square sets.

September 21, at Creede, while engaged as a foreman, Ray Morgan, age 39, was caught in a stope by a fall of rock and later died of his injuries. The mine where this accident occurred is operated through a cross-cut tunnel and by shrinkage stopes, and is timbered with stulls.

September 26, at Telluride, Enrico Zeber, a mucker, age 29, was struck by a rock that fell from a chute in the back of a drift, and fractured his skull. He died later.

October 30, at Aspen, while engaged as a timberman's helper, Rocco Chiaro, age 37, was fatally injured by a cave from the side of a stope. The mine where this accident occurred is operated through a shaft and by breast stopes and is timbered with square sets.

November 30, in the Leadville District, Joseph Karaza, age 33, a miner, while engaged in capping fuse, exploded 100 5X caps, which caused injuries that resulted in his death.

December 5, in the Cripple Creek District, while operating a machine drill, Seth Thomas Bradfield, age 40, was caught by a fall of rock and instantly killed. The mine where this accident occurred is operated through a shaft and by shrinkage stopes and is timbered with stulls.

December 21, at Telluride, while going down a raise, Sisto Sanchez, age 25, a mucker, lost his footing and fell 500 feet to his death. The mine where this accident occurred is operated through a tunnel, raises and by waste filled stopes and is timbered with stulls and drift sets.

December 24, at Georgetown, while employed in an upraise, J. C. Iverson, age 38, a timberman, was overcome by powder gas and fell 104 feet through a manway that was not constructed in an approved manner.

UNDERGROUND—1918

At Vanadium, January 18, while mucking, Bert Caughey, age 52, was almost instantly killed by a slab of rock that weighed several tons and fell from the roof. The mine where this accident occurred is worked through a tunnel and by breast stopes and the roof supported by posts, cribs and pillars.

February 2, in the Cripple Creek District, Charles Olson, age 50, was running a stoping machine, when the sprags that supported the staging on which he was standing gave way and he was precipitated 96 feet to the level below and killed instantly. The mine where this accident occurred is worked through a shaft and by shrinkage stopes and timbered with stulls. An attempt was being made to recover a block of ore after the stope had been drawn.

February 23, at Leadville, while cleaning out a chute, George Baxter, age 27, was struck on the head and instantly killed by a slab of rock that fell from the roof. The mine where this accident occurred is worked through a shaft and by breast stopes and timbered with square sets.

February 26 at Leadville, while engaged removing lagging between a manway and an ore chute, Joseph Kure, age 31, was struck on the head by some falling object and his skull fractured. He had evidently leaned into the ore chute to recover a pick that he had dropped. The mine where this accident occurred is worked through a shaft and by breast stopes and is timbered with square sets.

March 19, at Leadville, O. E. Dillon, age 35, was struck on the ribs by an iron bar which he was using to bar down loose rock, and sustained fatal injuries. The mine where this accident occurred is worked through a shaft and by breast stopes and timbered with square sets.

March 23, at Breckenridge, while descending a partly filled stope, Ben O. Sutton, age 21, dropped his hat and in his attempt to recover it, started a run of broken ore, that buried him and caused his death by suffocation. The mine where this accident occurred is worked through a cross-cut tunnel, an incline shaft and by shrinkage stopes and timbered with stulls.

March 29, at Romley, while going into a stope carrying a hose and an air drill, Richard B. Green, age 22, was caught by a fall of rock from the back, and crushed to death. The mine where this accident occurred is operated through a cross-cut tunnel, and raises, and by shrinkage stopes, and is timbered with stulls.

April 3, at Urad, while mucking, Otto Huseby, age 36, was instantly killed by a large cave of vein matter from the hanging wall. The mine where this accident occurred is worked through a tunnel, and by shrinkage and waste filled stopes and timbered with stulls.

April 12, at Lake City, while running a stoper drill, Tony Valenzich, age 45, was drawn into a filled stope and suffocated. This accident was caused by the withdrawal of ore at the level below, by his helper who was trying to provide more room for the operation of the drill. The mine where this accident occurred is worked through incline shaft and by shrinkage stopes, and timbered with stulls.

May 2, at Telluride, while cleaning up the bottom of the drift to make room for a mud sill, Oscar Sunberg, age 30, tunnel contractor, was caught by a run of dirt and fatally injured.

May 10, at Telluride, while tramming on a level, Ivan Raisch, age 28, was instantly killed by a rock that fell down a manway and struck him on the head. The mine where this accident occurred is operated through tunnels, raises and by waste filled stopes and is timbered with stulls and drift sets.

May 28, at Georgetown, while passing through a manway, Henry Smith, age 28, was struck on the head by a large rock and his skull fractured.

June 1, at Leadville, while shoveling in a new stope, John Gileovich, age 51, was instantly killed by a slab of rock that fell from a slip. The mine where this accident occurred is worked through a shaft and by breast stopes and timbered with square sets and cribs.

June 2, at Bonanza, while cleaning out the dirt for a set of timbers, William Applebee, a mucker, age 36, was crushed by a fall of rock from the back of the drift and died soon afterward. The mine where this accident occurred is worked through a shaft and by overhead stopes and timbered with drift sets, stulls and posts.

June 3, near Ouray, while shoveling in a stope, William Rebscher, age 48, undermined a large slab of rock on the hanging wall, which fell on him and caused instant death. The mine where this accident occurred is operated through a cross-cut tunnel by waste filled stopes and is timbered with drift sets and stulls.

June 4, at Cripple Creek, while barring down the walls of a stope, Harry Harvey, age 41, was struck on the head and instantly killed by a three-pound rock that had fallen about 15 feet. The mine where this accident occurred is worked through a shaft and by shrinkage and waste filled stopes and timbered with stulls.

June 6, near Silverton, while shoveling ore into a chute in an open stope, Frank White, age 17, was caught under a large slab of rock that fell from the hanging wall and instantly killed. This stope was open overhead for about 100 feet. The mine where this accident occurred is operated through adit drifts and by shrinkage stopes and timbered with stulls.

July 2, at Leadville, William Thomas Comfort, age 33, foreman miner, died from powder gas. A machine round had been blasted the previous day in a drift about 65 feet from the shaft and 190 feet from the surface. After loading one bucket he signaled the engineer to hoist, but when the bucket was returned to the bottom of the shaft the engineer received no response and he became alarmed. An investigation was made and Comfort's body was found face downward in the drift.

July 3, near La Plata City, while assisting in blasting a round of 14 machine holes in a wet cross-cut tunnel, Glenn Funk, age 23, was instantly killed when the first shot exploded. He had remained at the breast too long after spitting the first fuse.

July 27, at Telluride, Joseph Jaromillo, a machine man, was instantly killed by a fall of rock from the back of the stope. This mine is worked by an adit tunnel and by waste filled stopes, and is timbered with stulls and drift sets.

July 31, near Nederland, Joseph Aschwanden, age 42, a lessee, was found dead in a mine where he and his partners were driving a drift with machine drills from the bottom of a 95-foot shaft. They had blasted a round of holes in the afternoon, and Aschwanden, who lived alone in a cabin near the mine, evidently went into the mine in the evening to investigate the result of the last round, and was there overcome by powder gas.

August 5, at Telluride, while assisting to timber a manway, Chas. Poezi, a timberman's helper, was instantly killed by a fall of several tons of rock. The mine where the accident occurred is operated through a tunnel and by shrinkage stopes and timbered with stulls.

August 24, at Telluride, while drilling with a stoping machine in loose ground, Antonio S. Haspaas, age 33, was instantly killed by a fall of rock that was estimated to be several tons. This mine is worked through a tunnel, waste filled stopes, and timbered with stulls.

September 27, at Romley, while shoveling in a stope, Peter Verst, age 24, was instantly killed by a fall of vein matter from the back. The mine where this accident occurred is worked through a tunnel, by shrinkage and waste filled stopes and is timbered with stulls.

September 28, at Telluride, while going down a manway to turn off the air, Steven Slickovich, age 35, a machine man, fell a distance of about 55 feet and was instantly killed. The mine where this accident occurred is operated through a tunnel and by waste filled stopes and is timbered with stulls.

October 8, at Ironton, while cleaning out for a set of timbers, Dan Tedrezzi, age 39, a timberman, was instantly killed by a fall of rock from the side of a drift. The mine where this accident occurred is only a prospect.

October 13, at Georgetown, while working in an upraise, John Carlson, age 55, was overcome with gas and fell 125 feet to his death. The mine where this accident occurred is operated through a tunnel and upraises and by filled stopes and is timbered with square sets and stulls.

October 14, at Leadville, while cleaning out for a square set, William Walsh, a timberman's helper, age 18, was instantly killed by falling rocks. The mine where this accident occurred is worked through a shaft, and by breast stopes and is timbered with cribs and square sets.

November 9, at Ouray, while shoveling ore into a mill hole, E. Quintanos, age 25, was instantly killed by a fall of rock from the back of the stope. The mine where this accident occurred is operated through a tunnel, by both shrinkage and waste filled stopes, and is timbered with stulls.

November 23, at Telluride, Claude T. Denny, age 45, was instantly killed by falling 430 feet down a raise that was equipped as a hoistway. This raise was guarded by rails and covered with a trap door, and as no one witnessed the accident, it is not known how it occurred. The mine is operated through a tunnel and by shrinkage stopes, and is timbered with stulls.

December 21, at Leadville, while unloading timbers at the bottom of a raise, Matt Plutt, age 32, was instantly killed by a timber that fell down the raise and crushed his skull. The mine where this accident occurred is operated through a cross-cut tunnel, and underground shaft, and by breast stopes and is timbered with square sets.

IN SHAFTS—1917.

January 25, at Georgetown, while engaged as mucker in bottom of shaft, Nels Neilson, age about 35, was struck on the head by a rock which fell from above and fractured his skull. This caused his death five days later. The rock is supposed to have fallen from an overloaded bucket. The mine where this accident occurred is operated through a cross-cut tunnel, raises and winzes, and by waste filled stopes, and is timbered with drift sets, cribbing and stulls.

February 4, in the Cripple Creek District, while engaged sinking a shaft, William I. J. Langdon, age 47, was struck on the head and instantly killed by a rock that fell from a loaded bucket that was being hoisted.

On February 7, at Leadville, Joseph Triager, age 34, and Peter Matson, age 35, both miners, were standing on a cage, chopping ice from the side of a shaft 200 feet below the collar. The topmen discovered that the cable was out of the shieve wheel, and ordered the engineer to hoist the cage to the surface. When the power was applied the cable parted at the shieve wheel, allowing the cage to drop to the bottom (600 feet), killing both men instantly. The cage was equipped with safeties, but they failed to catch on account of the ice on the guides.

March 7, at Leadville, while engaged in mucking in the bottom of the shaft, Anton Miller, age 43, was struck on the head and killed by a rock that fell from a bucket as it was being hoisted.

March 9, in the Cripple Creek District, while engaged as a cager, Robert T. Norton, age 23, was killed by a mine car which was accidentally pushed into the shaft on the surface.

March 26, at Telluride, Arthur B. Lindgren, age 19, a compressor man, while riding in a winze skip, was caught by a large splinter (formed when the lip of the skip plowed into a ladder) and disemboweled.

June 5, at Leadville, while engaged as a cager, Angelo Passari, age 27, was caught between the bottom of the cage and the wall plate of the first set of timbers in the back of a station and was instantly killed.

June 16, in Cripple Creek District, while engaged in putting a bulk head in a shaft, John W. Dinwiddie, age 65, a mine carpenter, was caught between the shaft timbers and cage and crushed to death.

July 16, at Westcliffe, Joseph Garner, age 56, a timberman, fell 165 feet from a bucket which was being hoisted in a shaft and killed instantly.

August 18, at Idaho Springs, while engaged as a trammer, Arthur Povinelli, age 27, pushed a car loaded with ore into an open shaft, and was killed when he fell with it to the bottom.

November 1, in Cripple Creek District, while unloading timbers, James M. Blair, age 29, a substitute cager, fell between the shaft timbers and the cage and was crushed, when the wrong signal was given the engineer.

December 17, in Cripple Creek District, Fred Southerland, age 32, a shift boss, while descending a shaft on a bucket with another man, was struck by a floating cross-head and instantly killed.

December 28, in Cripple Creek District, while mucking in the bottom of a shaft, Charles J. Peterson, age 45, a timberman, was struck on the head and killed by a rock that fell from a bucket as it was being hoisted to the surface.

IN SHAFTS—1918

April 1, at Leadville, while engaged in repairing the shaft timbers, Robert Adamson, age 40, was thrown from the cage (when the staging planks with which he was riding up the shaft caught under the shaft timbers) and crushed to death under a wall plate.

Between April 1 and April 10, at Central City, while repairing an old shaft, Charles Anderson, age 60, lost his footing and fell about 100 feet to the bottom of the shaft, where he was found several days afterward. He had evidently been killed instantly.

August 27, in the Cripple Creek District, while being hoisted in a bucket, C. L. Bembry, age 27, was struck on the head by a piece of 1-inch air pipe which had fallen from the level above and instantly killed.

SURFACE PLANTS AND SHOPS—1917

May 25, at Leadville, John O'Neill, age 39, a timberman, was helping place props under the sloping side of a surface ore bin, when the bottom of the bin gave way, and he was caught by a rush of ore that injured him fatally.

July 26, at Leadville, while stooping down to adz off a door sill, Frank Seabrey, age 43, a watchman, was shot through the thigh by a gun which had dropped from his shoulder holsters. He died from his injuries a few weeks later.

October 13, at Vanadium, while hoisting timber with a pulley and guy rope, George W. Miller, age 63, a caretaker, was struck by the swinging timber and received injuries which subsequently proved fatal.

SURFACE PLANTS AND SHOPS—1918

April 12, near Nederland, John C. Downtain, age 82, was caught by a cave of ground from the face of an open cut and smothered to death.

July 8, at Lakewood, Paul Froelich, age 58, was struck by a cave of loose dirt from the face of an open cut, and his neck broken.

August 3, on the mountain above Telluride, while engaged making a survey, Guy Erickson, civil engineer, age 29, lost his footing, fell from a small cliff, and received injuries which resulted in his death.

May 29, at Lakewood, while testing in a transformer building, John A. Delmar, age 34, an electrician, took hold of a high tension line and was electrocuted.

November 19, in the Cripple Creek District, while pulling sheet iron off the roof of a mill building, Emil Palmquist, age 25, a laborer, touched a steel bar against wires carrying a current of 23,000 volts, and was electrocuted.

ORE DRESSING AND MILLING PLANTS—1917

January 10, at Boulder, Thomas Coutts, age 27, a mill man, attempted to move rocks from the crushing rolls and had his hand pulled in and torn off. He died later.

February 26, at Boulder, Maxwell Duff, age 21, a mill man, while putting dressing on a belt that was slipping, was thrown against a set of unguarded crushing rolls, when the belt flew off the pulley. He became entangled in the collars of the rolls, and died from his injuries.

May 17, at Red Cliff, Ike Shreve, age 51, a hopper man, while walking on a temporary platform over ore coolers, stepped on a light board, that broke, precipitating him to the floor below, where he sustained injuries which caused his death seven months later.

June 9, at Boulder, Isadore Caserta, age 34, a mill roustabout, was sent to paint the roof of a transformer house, and was electrocuted by that carried high voltage and passed close to the roof.

June 13, at Aspen, while winding the bottom dump door of a railroad car, Thomas Stringer, age 52, a mill man, was either struck on the head with the winding lever or else was thrown so that his head struck the railroad rail, and killed.

July 17, at Georgetown, while throwing a three-pole electric switch, Harry Morris, age 19, a mill man, touched one of the poles of the switch and was electrocuted.

August 20, at Silverton, Victor Sapp, age 35, a tramway grip-man, in attempting to get off a moving aerial tramway bucket at a tower lost his balance, fell to the cliffs below and was killed instantly.

ORE DRESSING AND MILLING PLANTS—1918

June 2, at Telluride, A. J. Sneddon, age 33, a mill man, was attempting to put a belt on a running pulley, when his clothes caught on a set-screw on a revolving shaft, and he was wound around with the shaft and beaten to death.

December 3, in the Cripple Creek District, while engaged as a substitute motorman, M. F. Perdue, age 25, lost control of an electric-driven ore car and was thrown into the ore bins and covered up with ore, so that he received injuries which caused his death.

SMELTERS—1917

February 4, at Salida, while unloading coal, Julius Murton, age 51, a laborer, was either struck on the head or thrown off his balance by the lever of a dump car and received injuries that caused his death.

SMELTERS—1918

March 6, at Florence, while attempting to get on to a moving freight elevator, John I. Velarde, age 20, a furnace feeder, was caught between the elevator frame and a steel girder, where he received injuries that caused his death.

March 14, at Salida, while cleaning the bottom of a slag pit, Antonio Gregovich, age 36, laborer, was caught by a runaway slag car and instantly killed.

June 9, at Boulder, while engaged in placing corrugated iron on roof of small building near a transformer house, Glenn McFall, age 21, came in contact with a high tension line and was electrocuted.

July 23, at Pueblo, while in a pit, Reyes Neito, age 20, a laborer, came in contact with a revolving shaft so that his clothes were wrapped around it, and he received injuries that proved fatal.

October 20, at Leadville, while standing in front of a motor, Glenn Bartlussi, age 15, a motorman, replaced the trolley pole that had jumped off the line, which caused the motor to surge forward over his body, and injured him fatally.

ROCK QUARRIES—1917

October 25, near Pueblo, while hauling stone from a quarry to the mill, Harry Bolster, age 26, a teamster, was thrown from the top of the load by the slipping of the load of stone so that one of the slabs fell on him and caused his instant death.

ROCK QUARRIES—1918

May 18, at Thomasville, while sitting on a cap pulling up timbers, Mannie Deeds lost his balance, fell on his head, and received injuries that caused his death.

DREDGES—1917

November 21, at Breckenridge, while receiving coal in a scow, Harold John Roberts, age 18, a laborer, was thrown into the pond and drowned, when the scow on which he was receiving the coal was capsized by a cave of gravel from the bank.

REVIEW OF DISTRICT No. 1

By INSPECTOR T. R. HENAHEN

There was not as much activity in the mining industry in the First District during the year of 1918 as there should have been. The scarcity of labor, advance in the cost of mining supplies, and the increase in freight rates, smelting charges and wages, all had some effect on the output from the mines. The gold and silver producers of America were the real patriots during this war period, for they continued to operate even though they received a fixed price for their product, while every other industry in the country was taken care of by the Government.

GILPIN COUNTY

Gilpin, while the smallest in area is one of the most important counties of the state. It was organized in 1861 and named in honor of Colorado's first Territorial Governor. As now constituted, it has an area of 185 square miles and occupies a north central position in the state. Considering its area, Gilpin County has been one of the heaviest producers of gold on this continent.

During the winter of 1917 and 1918, the Perigo and War Eagle mines located at Perigo were cleaned out and retimbered and the mill overhauled and put in operation.

The Pine Mining District, six miles north of Central City, is where the Evergreen Mines Company properties are located. During the fall and winter of 1917 and 1918, they erected a 100-ton flotation mill, which is equipped with electric power. This company gives employment to a great number of men the year round. The Pennsylvania and Colorado Mining Tunnel and Milling Company's mines and mills are located in this district and several other smaller properties.

In the Silver Creek District, five miles northwest of Central City, the Cornucopia mine and mill are located. This property is being thoroughly developed so that it can produce a heavy tonnage during the coming year.

In the Russell Gulch Mining District there are in operation such properties as the Becky Sharp, Frontenac, Fairfield, Forfar, Jupiter, St. Clair and several other smaller ones.

In the Willis Gulch Mining District, the following mines are in operation: Gilipado (better known as the Chase mine), Powers mine, Two-forty mine, Hampton mine and several smaller ones.

In the Quartz Hill and Nevadaville mining district the following mines are in operation: The Denver, Eureka, Roderick Dhu and several other small properties.

In the Black Hawk District are the Redick Mining Company's properties (better known as the Bryan Tunnel), and the Star Mining & Tunnel Company's properties and several other smaller ones.

CLEAR CREEK COUNTY

Clear Creek is one of the most important counties in the state; it was organized in 1861 and bears the distinction of being the scene of the discovery of the first pay placer beds in the state. From 1859—the year of this discovery—at the mouth of Chicago Creek up to the present day, mining has been continually prosecuted and each year productive of important development

The Daly Mining District (11 miles west of Empire) is where the Primos Exploration Company's mines and mills are located. Late in the fall and winter of 1917-1918, this company erected a 150-ton mill for the treatment of their molybdenum ores. This is one of the largest mines of its character in Colorado, giving steady employment to 125 or 150 men.

The Silver Mountain Mining District, north of the Town of Empire, is where the Golden Empire Mining Company's properties are located. One-fourth of a mile west of Empire are the Twin Ports and Empire Mining Company's properties. There are several other smaller properties being developed in a small way in this district.

In the city limits of Georgetown, the Onondago mine, Centennial mine, Capital Mining & Tunnel Company mines and mills and the Georgetown Tunnel & Transportation Co. and several others are in operation.

The Argentine Mining District, about 1½ miles west of Georgetown, is where the Colorado Central Mining Company's mines and mill are located. During the fall of 1917 and the winter of 1918, this company erected a 500-ton flotation mill. In the upper Argentine Mining District, 8 miles northwest of Georgetown, is where the Santiago Consolidated Mining, Milling and Tunnel Company's mines are located. This mine is being operated by the company and certain parts of the mine are operated under lease. The Kitty Ousley mine is also located in this district, besides several others.

In the Silver Plume District, or part way between Silver Plume and Georgetown, are the Wide West Mining Company's properties. This property is better known as the Hall Tunnel.

At Silver Plume is located the Pay Rock, whose tunnels were cleaned out and retimbered last fall. One-half mile west of Silver Plume are the Wasatch Colorado Mining Company's mines and mills. This property is better known as the Mendota mine. Three-fourths of a mile west of Silver Plume is the Hollingsworth Mining Company's properties. This mine was known as the Smuggler mine. A short distance west of the Smuggler is the Denbeigh silver and lead mine and mill. This mine is known as The Terrible. About 1½ miles northwest of Silver Plume is located the 7:30 mine.

Six miles southwest of Silver Plume is located the Josephine mine and mill. Seven miles southwest of Silver Plume is located the Mount Kelso Mining Company's properties in the Grizzley Gulch and several other smaller properties in this district operate in a small way.

In the Lawson Mining District is the Bellview-Hudson Tunnel Company's mines. The Twin Sisters mine, The Jo Reynolds mine and mills, Matrix Mining Company, Teddy Bear Mine, Princess Tunnel and the Little Giant Gold Mining & Milling Company's mines and mills.

In the Dumont Mining District, The United Freeland Development and Tunnel Company's properties, the Blue Ridge and the Senator mines and mills, the Albro mine and Albro Extension and others, were in operation.

Freeland Mining District: There are in operation The Broderick Mining & Milling Company's mines and mills. This mine is known as the Lamartiwe mine. In the town of Freeland are the New Era mines and mill, Turner mine, Pine Tree and several others.

In the Alice Mining District is The Gold Anchor mine and mill, the Klondyke Tunnel Company and The Lombard mines and mill.

In the Virginia Canon Mining District, The Two Brothers Mine, The Belleview Champion mine and the Crown Point mine, which was cleaned out and retimbered in the spring of 1918. There are several other smaller properties in operation.

In the Gilson Gulch Mining District there are The Silver Edge Mine, Franklin Mine, Sun and Moon Mine and Cincinnati. All operated under the leasing system.

In the Chicago Creek Mining District there are the Burns-Moore Tunnel Co. Silver Horn Mine, Mattie Mine and others.

In the Upper Idaho Springs, or what is known as the Spanish Bar District, are located the Big Five Tunnel Co.'s properties, which are operated under the leasing system. West of the Big Five is the Stanley mine. This property is operated under the leasing system.

What might be called the Idaho Springs Mining District, including the Newhouse Tunnel and properties, operated through the tunnel, such as the Gem mines. There are about ten leasing companies operating in the different parts of the properties owned by the Gem Company. The Golden Edge Mining Co. is also operating through this tunnel. The Sun and Moon Leasing Co., working in the lower levels, is also operating through the tunnel. The Jenkins & Johnson Leasing Co. are operating the Pozo mine through this tunnel. The apex of this vein is in Gilpin County. The Tremont Mining Co. is also operating through the tunnel; the apex of their vein is also in Gilpin County.

At the portal of the Newhouse Tunnel is a 200-ton mill, thoroughly equipped for handling all kinds of ore.

BOULDER COUNTY

Boulder is one of the pioneer counties and has been an important factor in the state's history. While not first to discover gold in paying quantities, it was first to make the discovery known. It bears the distinction of being the scene of the first general excitement following the Arapahoe County fiasco; of organizing the first mining district and formulating a code of laws for local government; of locating the first quartz vein, erecting the first stamp mill and having the first steam sawmill; building the first board house and the first school house; constructing the first irrigating ditch for agricultural purposes, and the first blast furnace for the manufacture of iron.

During the past year the tungsten producers have not been receiving what they should have received for their product on account of the unsettled conditions in the market. Just as soon as things adjust themselves and conditions become normal again, I can see no reason why the tungsten industry should not become active again.

Some of the largest producing companies in Boulder County are the Primos Mining & Milling Co., Boulder Tungsten Production Co., Wolf Tongue Co., Tungsten Products Co., Vasco Mining Co., Tungsten Metals Corporation, Long Chance Mining Co., The Mojave Boulder Tungsten Mining Co. and dozens of others too numerous to mention.

The fluorspar industry in the Jamestown Mining District was very prosperous during the war period and gave employment to a great number of men. The largest producing mines were the Blue Jay, Argo, Chancellor, Wano, Emmett and dozens of others.

In the Caribou Mining District there has been a great revival in 1918. The mines in operation are the Caribou, Allied Gold Mining Co.'s mines, Belcher mine, and the Potosi and several other smaller ones.

In the Ward Mining District there has been considerable activity in 1917 and 1918. The Staughton reopened and the Utica mines were unwatered and retimbered.

The White Raven mine, which is located about 1½ miles southeast of Ward, is one of the best small silver mines in northern Colorado.

In the Gold Hill Mining District are the Slide mines and mill. The reopening of the Horsfall mine has put new life into Gold Hill.

The Golden Age Mine and Mill are located about 2½ miles southeast of Jamestown.

GRAND COUNTY

History reveals the fact that this section was entered by the white man in 1859, and also that it was a favored spot with the Indians, who were loath to give up this sheltered domain. While this section has been the scene of several mining excitements, its inaccessibility has proven disastrous to development. The county is prolific in mineral resources, but they are almost entirely undeveloped.

The Mary Florence Mining, Milling and Leasing Co.'s property is located 12 miles south of Parshall, a station on the Moffat Railroad. This mine has been known as the Mollie Groves.

The Ready Cash mine is located at the head of Williams Fork in the La Plata Mining District. This mine has been operated in a small way for the past 30 years and has produced a great amount of high-grade silver ores.

The Co-Operative Metal Mines Co.'s properties are located in the La Plata Mining District at Williams Fork. This mine is also being pretty thoroughly developed. There are several others prospecting in this district.

ROUTT COUNTY

Hahn's Peak Gold Mining Co.'s mines and mills are located at the foot of Hahn's Peak. They got their mill in operation last fall at the time the cold weather set in and had to suspend operations during the winter months.

There is also some gold dredging and placer mining done in and around Hahn's Peak.

MOFFAT COUNTY

The Douglas Mountain Copper Mines Co. are located 31 miles northwest of Sunbeam. This company is also erecting a mill on the ground.

JACKSON COUNTY

The Wolverine Mines & Mill are located in the Wolverine Mining District, 32 miles northwest of Walden. The Village Belle Mine is located 19 miles northwest of Walden and 4 miles west of Kings Canon, a station on the Colorado, Wyoming and Eastern Railroad. The North Park Mining and Milling Company properties are located in the Teller Mining District, 10 miles southeast of Rand and 35 miles southeast of Walden, the county seat of Jackson County.

JEFFERSON COUNTY

One of the principal industries of the county is the manufacture of clays. These beds, while common to the Front Range border counties, are unusually developed in this region and of superior quality. The fire brick, press brick, tiling, pottery, sewer pipe, etc., manufactured from the Jefferson County products, lead that of any other part of the state.

The largest companies producing clay in Jefferson County are Ruby-Parfet & Co., The Denver Fire Clay Co., and The Denver Fire Brick Co., J. T. Williams, Golightly and Bennett.

DENVER COUNTY

Here is where the American Smelting & Refining Co.'s plant is located, the Western Chemical Mfg. Co., the Chemical Products Co., the Colorado Radium Co., The Denver Fire Brick, The Denver Fire Clay, and dozens of other manufacturing concerns, too numerous to mention.

In conclusion will say that, owing to the Federal laws on explosives and other laws, we were compelled to make a great many verbal recommendations. I made 108 verbal recommendations to mine operators during the past year.

REVIEW OF DISTRICT No. 2

By INSPECTOR M. J. McCARTHY

TELLER COUNTY

While the ore production for the Cripple Creek District has been somewhat curtailed for the past year, this is accounted for by insufficient labor, with increased wages, and the tremendous increase in the price of powder and all other mining supplies, with, of course, no relative increase in the price of gold. This has had a natural tendency to slow down development work, with the mines working only on their developed reserves.

During the year a great many of our miners have left for camps whose products were so affected by war prices that they were able, temporarily, to pay higher wages which looked very attractive to the miner. A great many men were also drawn into the service of the United States, which fact added to the already serious shortage of labor.

Now that peace is assured, with the probability that the price of powder and all other mining supplies will soon be back to normal, and with our men released from the army in great numbers, we can feel sure of a great revival in the mining interests of the Cripple Creek District for the coming years. Development work will be greatly augmented and new ore reserves opened up, leasers will be operating in greater numbers, and in general the future of the district looks more promising than for several years past.

Great credit is due the mine owners for their patriotism in keeping their mines in operation under such adverse conditions, when the same ore, left in the mines to be taken out at some future time and under normal conditions, would have netted them a very handsome saving.

The Eagle Ore Company, samplers and ore buyers. The business of this company has been very regular during 1917 and 1918. Though a little below the average of former years, nevertheless the company has handled an average of 8,500 tons per month of good grade ore during this biennial period.

The Gold Sovereign Leasing Company. During 1917 and 1918 this company has done a little over 300 feet of drifting and cross-cutting and sunk the main shaft 150 feet. But little work was done during the year 1918. It cost better than $13,700 to do the development work, and about 500 tons of medium grade ore were shipped.

The Victory Gold Mining Company operated the Prince Albert property on Beacon Hill under bond and lease. Fifteen cars of medium grade ore were shipped, and about the middle of 1918 the company discontinued operations with the exception of a little subleasing.

In March this company acquired the Smith lease on the Howard shaft of the Mary McKinney Mining Company and have operated it continuously ever since. A good production of fair grade ore is being made from the property. There are also several sets of sublessees operating on the split check system.

It is expected that a heavy production will be made during the coming year. This company also owns a considerable acreage on Ironclad Hill and a little prospecting was done recently for manganese.

The Erie Mines Company. Very little work has been done on the properties of this company during the past two years. A few cars of $15.00 ore were shipped from the Rubie mine, but the company is now inactive.

The Cresson Consolidated Gold Mining and Milling Company. The production of this property during 1917-1918 was 189,078 tons. The net returns were $3,987,851.00. The company received as additional income $7,526.00 net royalties from 5,083 tons of ore shipped by lessees.

The development consisted of 7,385 ft. diamond drill holes, 10,485 ft. drifts and crosscuts, 2,471 ft. raises and winzes, 192 ft. sinking main shaft.

The most interesting features resulting from the two years' operations are:

Discovery of new ore bodies on the third, fourth and fifth levels on the vein system of the Funeral and Silver dykes.

Discovery of new ore bodies of very substantial size and value on the seventh, eighth, ninth and tenth levels, located in the center of the volcanic crater. The past development and ore discoveries in the eruptive area have all been limited to the contact or exterior boundaries of the crater.

Discovery on the Roosevelt Tunnel level of ore bodies on the downward extension and increase in size and value of the ore measures heretofore disclosed on the sixteenth, or deepest level of the Cresson, 175 feet above the tunnel.

From all present indications and the new discovery of so many large ore bodies within the last year, the mine has a long, bright and prosperous future.

The connection made from the Roosevelt Drainage Tunnel with the bottom of the Cresson shaft affords much better ventilation on all lower levels of the mine, and when the connection is made between the tunnel and the winze which is located near the center of the main ore bodies on level sixteen, it is bound to be of great advantage to the company as it will improve the air conditions and labor efficiency very materially.

The Forest Queen Mines Company has made very good progress since starting operations two and a half years ago, and found it necessary to enlarge and repair the shaft. After March 1, the shaft will be sunk 200 feet deeper.

During the period under review, 700 ft. of crosscuts, 120 ft. of upraises and 100 ft. of shaft work was done and the production of 3,575 tons of fair grade ore was mostly shipped to the Golden Cycle Mill, at Colorado Springs.

The Portland Gold Mining Company, for the years 1917-1918, performed development work as follows: 18,600 ft. drifts and crosscuts, 3,150 ft. raises and winzes, 140 ft. sinking No. 2 shaft, 460 ft. sinking surface shaft, 2,230 ft. work on Roosevelt Drainage Tunnel.

The total amount of development done on the property now amounts to about 80 miles.

During the year 1917, 88,625 tons of high-grade ore of a gross value of $1,768,486.27 was shipped and 469,877 tons of low grade valued at $961,153.90 has been shipped to the company's mills, making a total tonnage for the year of 556,502 of a gross value of $2,729,640.17.

The connection with the Roosevelt Drainage Tunnel at a depth of 2,133 feet is very satisfactory and the prospects for large ore bodies on this level are very encouraging, and the probabilities are favorable that a large and up-to-date electric pumping plant will be installed here for a further prosecution of deeper development.

Portland Mill (Victor). The Colorado Springs Mill of the Portland Gold Mining Company was closed April 1, 1918, and the Victor Mill on top of Battle Mountain was closed August 1, 1918.

The low grade milling of the Portland Gold Mining Company is now centralized in one plant, "The Independence Mill," which has a daily capacity of 1,350 tons and is located on the old site of Argall's "Stratton's Independence Plant."

The Cripple Creek Drainage and Tunnel Company. No work has been done in the breast of the main tunnel since July 1, 1918. The total length of the tunnel on January 1, 1919, was 23,937 ft. Length of crosscut from tunnel connecting Cresson shaft, 1,715 ft.; length of crosscut from tunnel connecting Portland No. 2 shaft, 2,000 ft.

The depth of shafts which are connected with the drainage tunnel are as follows: El Paso No. 1, 289, ft.; Elkton, 1,640 ft.; Cresson, 1,915 ft.; Portland No. 2, 2,133 ft. The depths of these shafts are from the surface to the Drainage Tunnel level.

The Mary McKinney Mining Company have done a great amount of development during this biennial period. This company has had at various times as many as eight sets of lessees working on the property, with an average of four all the time.

The Vindicator Consolidated Gold Mining Company employs approximately 266 men on their properties. Development work during the major part of the past year has been practically discontinued.

The fact that the company has done a total of 3,640 feet of development work on the lowest levels of the mine, thoroughly prospecting the known ore zones without proving up any great amount of commercial ore and the acute shortage of labor, caused them to abandon the lower levels of the mine and discontinue pumping from this area.

The tonnage produced for the years 1917-1918 will be about 467,000 tons, valued at approximately $2,743,000.

An average of about 20,000 tons of ore is treated at the company's mill, on which a profit of about 35 cents per ton is made.

The Free Coinage Gold Mining Company. The property of the Free Coinage Company has been worked at intervals by lessees during the past two years, but the production has been light. The property is inactive at present but may begin operating early in 1919.

The Gold Pinnacle Mining Company. Some work was done on the Mitchell shaft on the eastern slope of Bull Hill by lessees, light tonnage of dump ore being shipped. The property is now inactive.

Block 8 of School Section 16. Operations ceased on Block 8 during 1917, and with the death of Alfred La Montaigne the lease has remained inactive except for a light shipment of dump ore.

Anderson and Benkleman, lessees, secured a lease on the Dexter Mine in 1909 and operated the property three years before opening up ore to any extent, spending about $30,000.00 in development work in that time. In 1912 they were rewarded by a rich discovery on the 1,000-foot level and have continued to be heavy shippers ever since.

In January, 1915, they secured a lease on the Trail claim of The United Gold Mines Company and within two months were shipping ore of a good grade, working through the Dexter shaft.

In order to handle the ore efficiently and economically from both properties, a complete plant of machinery was installed on the Dexter shaft, consisting of an electric hoist and compressor and an air drill sharpener. They also sunk the shaft from a depth of 600 feet to 1,400 feet and have done over 10,000 feet of drifting and crosscutting to thoroughly explore these properties.

Their principal operations for the last two years have been on the lower levels, from the 11th to the 14th. In that period they have shipped 1,124 cars of ore from the Trail and 173 cars from the Dexter, averaging 30 tons to the car and of value better than $20.00 per ton. At this time they are mining a good grade of ore on level 14 in the Dexter property and will sink the shaft at least one hundred feet deeper in 1919.

The Colburn Ajax Mine is being operated very successfully by The Carolina Company, a Massachusetts corporation, under lease and bond.

The production for 1917-1918 was 18,760 tons and the gross value $405,250.00.

Development: On levels 14, 16, 18 and 20, drifts and crosscuts 8,194 feet; sinking the main shaft 502 feet, which gives this shaft a total depth of 1,981 feet.

The Queen Bess Mining Company. During the last biennial period this company has received net returns from ore amounting to $109,374.72.

Development: On the Queen Bess property, 2,800 feet; on the Gold King property, 1.250 feet; and stoped the Queen Bess vein vertically 550 feet.

The Lincoln Mines and Reduction Company acquired the property formerly held by the Rex Mining and Milling Company in July, 1918. This is a 35-acre tract on Ironclad Hill in the Cripple Creek District consisting of the following claims: Ironclad, Magna Charta, Quartzite, Pard and Annex.

The Lincoln Mines and Reduction Company also owns the Lincoln Mine near Gillett, Colorado, and a considerable amount of property connected therewith. Such development work as had been applied to the Lincoln Mine previous to the war indicated gold ore to some value in depth, of which advantage will be taken as the business of the company develops.

Early in 1918 the company obtained a license for the use of a new process of ore concentration invented and patented by Mr. F. G. Gasche of Chicago, Illinois. The original intention of the company was to erect a mill and concentration plant at the Lincoln Mine, but development costs indicated the desirability of acquiring a new property within the Cripple Creek District, where the condition of ore deposits, availability of labor, mill sites and other factors favored an earlier development of the new concentration process.

While it was known that there were outcroppings of manganese ore on the newly acquired property of the company, a limited amount of prospecting and development revealed manganese ore in considerable quantity.

The mine development has scarcely started in view of the necessity of an - early completion of the mills and concentration plant.

The Stratton Cripple Creek Mining and Development Company, owning the largest acreage of any company in the district and located in the center of the camp, have operated their property for the past fifteen years by the leasing system, which had proved very successful up to the time of the beginning of the war. Since that time there has been a falling off of the number of leasers, and those remaining could not afford to do the proper amount of development work to keep up the average production they have had in the past. The development work done during the biennial period is as follows: drifts and crosscuts, 6,107 feet; raises and winzes, 2,748 feet; shafts, 1,005 feet.

Some of the most important development was the sinking of a good working shaft on the Longfellow to a depth of 600 feet and drifting to one of the large ore shoots on the 5th level. Also the sinking of the winze on the Baskett and Luce lease, on the ore below the 1,500-foot level, on the American Eagles Mine, which is located on Bull Hill.

The Acacia Gold Mining Company. The South Burns shaft has been completed to a depth of 1,460 feet and shipments of a very good grade of ore were regular during 1917 and up to November 30, 1918. The lessees have also been shipping a good average grade of ore. The ore opened up in the winze on the 14th or lower level is of a better grade, but in order to mine the ore economically the company have decided to sink the main shaft another lift of 100 feet; but have also decided to discontinue all work for the present and until such time as a fair reduction is made in the cost of mining supplies.

The Excelsior Mining, Milling and Electric Company, leasing the Longfellow claim of the Stratton Cripple Creek Mining and Development Company, operated during 1917 through the 5th and 6th levels of the Golden Cycle shaft. They have sunk the Longfellow shaft to a depth of 600 feet and made connection at 500 ft. with the 5th level of the Golden Cycle shaft, at which point a large body of fair grade ore was found. A crosscut has been started on level 6 to drive for a known ore shoot, which is a continuation of the ore shoot on level 5. This shaft is equipped with a good steam plant and a new skip. The company have also built a large ore house within the last year.

The Modoc Consolidated Mines Company was incorporated in 1916. The company took over the Modoc Mine, and at once directed its efforts to the sinking of a new vertical shaft to replace the old incline shaft which had reached a depth of 1,100 feet and through which the handling of ores from the lower levels had become costly. In May, 1918, a merger was effected whereby the Last Dollar Mine, an adjacent property, was consolidated with the Modoc Group. This group has a production record of approximately $5,000,000, the bulk of which was made from upper levels of the mine. Development during the past two years has totaled 1,889 feet of drifts crosscuts and raises, which is in addition to the work carried on in the new shaft. The Frankenberg shaft has been sunk 900 feet, the plan being to sink it to a depth of 1,800 feet.

'The recent consolidation with the Last Dollar Mine has provided another shaft which is 1,500 feet deep and from which the company will direct the exploration of the entire consolidated territory until the Frankenberg shaft shall have been completed. The new vein recently discovered on the 1,100-foot level is perhaps the most important to date for the company.

Work is now being directed on the 600-foot and 1,000-foot levels and they expect to cut the vein which was discovered on the 1,100-foot level. An ore body on the 500-foot level in the northern undeveloped area of the Last Dollar property shows a good grade of ore that is now being mined. The north end of the Last Dollar claim and the entire Combination claim are entirely undeveloped.

A new ore house is to be built in the spring, and material is now being delivered for its construction. The structure will be divided into six bins and will be equipped with washers, sorting belts, trommel screens, etc.

The Last Dollar shaft has been equipped with new machinery, consisting of an electric hoist adequate to operate to a depth of 2,000 feet, also a new compressor and a complete equipment for a blacksmith shop.

The Granite Gold Mining Company, owning a large acreage adjoining the Portland G. M. Company's property on the southern slope of Battle Mountain, have operated steadily during 1917 and 1918, under adverse conditions. The production amounted to over 3 000 tons per month of good grade of ore. Development consists of, drifts and crosscuts, 7,527 feet; winzes and raises, 1,300 feet; sinking Dillon shaft, 170 feet.

The Queen Gold Mining Company. This property was one of the first producers of the Cripple Creek District, and shipped at one time very rich ore. The present company, which is a new organization, has operated steadily during 1917 and until early in the summer of 1918, when the mine was closed down. The company had always maintained a steady production of good grade ore. The shaft was sunk to a depth of 1,015 feet.

The Cripple Creek Gold Mining Company, operating south of Victor avenue and west of the High School, have sunk the shaft in the Granite formation to a depth of about 1,000 feet.

The Ocean Wave Mining Company have operated steadily with little success, having made a production of about $2,400.

Development work consists of, drifts and crosscuts, 800 feet; winzes and raises, 100 feet; shaft sinking, 500 feet. Operations may be discontinued until such time as a reduction is made in the cost of mining supplies.

The Strong Gold Mining Company operated steadily with fair success until about November 1st, when operations were discontinued for the purpose of making necessary repairs to machinery and other equipment. When this is completed, work will be resumed. The main shaft has been sunk to a depth of 1,619 feet.

The Millasier Mining Corporation are the owners and operators of the Clyde Mine, which is favorably located, being adjacent to the Modoc and Portland Mines. The Clyde shaft was sunk from the 800-foot point to 1,400 feet in about seven months, and while the showing was very favorable the mine was closed down for the duration of the war. More than likely operations will start again some time in 1919.

The Gold Dollar Consolidated Mining Company. The Mable M. shaft on the eastern slope of Beacon Hill was operated under lease during 1917 and until February 22, 1918, by lessees who made a light production.

The property is now inactive, but with the new organization perfected, will again be operated on a much larger scale.

The Blue Flag Gold Mining Company. The Blue Flag shaft was sunk to a depth of about 1,200 feet before closing down the mine several months ago. The plant was steamed up once since then to permit the directors to examine the mine workings.

The Joe Dandy Mining Company. This property was operated during 1917 and part of 1918 by the Eclipse Leasing Company, who made a good production up to the time the company discontinned operations for the duration of the war. Operations are likely to start again within the next few weeks.

The Sheriff Gold Mining Company. This property was active for a few months during 1918, when the shaft was retimbered and considerable development work done.

The Moose Gold Mining Company. Some prospecting was done in the surface workings, and a light tonnage of dump ore was shipped.

At the Bertha B., owned by Carl Johnson, of Denver, several hundred feet of development work was done, a small vein of ore was discovered and a new ore house built, but no production was made. The property is favorably located, being about 1/4 of a mile from the famous Cresson mine.

The Petrel Gold Mining Company. This property is located on the southwestern slope of Squaw Mountain, and was operated for a few months, making light production.

The Elkton Consolidated Mining and Milling Company. The Elkton mine for the past two years has been used largely as a supply and operating shaft for the Roosevelt Deep Drainage Tunnel, with which it is connected. The property is also operated by several sets of lessees, who have made light production. The ore from the large ore bodies, opened up by the Cresson Consolidated Gold Mining and Milling Company at the tunnel level, has also been hoisted through this shaft.

The Palace Lode Mining Claim was operated on a small scale during 1917 and 1918 by Brandt and Grefig, with little success. There was no production.

The Ella W. This property was operated at different times during 1917 and 1918 by lessees who made light shipments of good milling ore.

On account of the proximity of large magazines for the storage of powder to the several mines and towns in the Cripple Creek District, it was advised that a central storage system be adopted, and safe sites selected. The several powder companies and mine operators were consulted and the plan agreed to, and early in 1918 The E. I. Du Pont De Nemours Powder Company built a storage magazine at a cost of $12,000.00, which has a capacity of 3,000 boxes, with a new railroad track and automobile road.

The Hercules Powder Company's magazine and automobile road was built at a cost of about $2,300, and that of the Atlas Powder Company, known locally as the Giant Powder Company, at a cost of about $2,100, which in each case includes the purchase price of the ground.

This plan has afforded the people of the entire Cripple Creek District the protection to which they were entitled.

The Index Gold Mining Company. This property is being operated by The El Paso Extension Gold Mining Corporation. During this biennial period they have sunk the main shaft 200 feet, completing it to a depth of about 1,100 feet. In addition to this they have done about 600 feet of drifting and cross-cutting on the upper levels, but now they are confining their efforts to the lowest level, where they are drifting and expect to open up large ore bodies on the two prominent fissures popularly known as the Pointer and Index veins. They have approximately 200 feet yet to drive before they may hope to encounter the ore bodies known to exist.

The Reva Mining Company is operating the property of the Rose Nicol Gold Mining Company under a long time lease. For the past two years all efforts were confined to development work.

The Rose Nicol shaft was sunk to a depth of 932 feet, from which point a drift was driven 900 feet, where a connection was made with the 10th level of the Portland No. 2 shaft. Other drifts were driven to connect with the Dexter Mine and new ore shoots have been opened up on several levels.

A' new ore house and an aerial tram extending from the mine to the loading station in Eclipse Gulch will be completed about the latter part of January, at which time the Reva Mining Company will start shipping from 3 to 5 cars of ore per week. The Company has started sinking another lift of 125 feet which will make the shaft about 1,100 feet deep.

The Masterpiece Mines Company. This property is located on the northeast slope of Big Bull Mountain, in a non-productive area. It is opened up by a tunnel and several thousand feet of development work, but no production has been made.

The Findley Mines Company, which has made heavy production in past years, has been inactive for two years, except for a few hundred feet of development work which was done near the line of the American Eagles mine, where a discovery of rich ore was made.

The Wilkshire Gold Mining Company's property on Raven Hill has been under lease to Porter Hedges, who made a small production during 1918.

The Tornado Mine, owned by the Elkton Consolidated Mining and Milling Company, has been operated by George E. Collins, who made a fair production. The property is now inactive.

The Hiawatha Gold Mining Company, located on the southwestern slope of Beacon Hill, was operated by lessees part of 1917 and 1918, who also operated on the El Paso property through the Hiawatha shaft. Light production was made of medium grade ore.

The Rocky Mountain Gold Mining and Milling Company. This property is located on the north slope of Beacon Hill. It was active under lease to the El Paso Extension Gold Mining Corporation, who operated for several months without success.

The Commonwealth Mining and Milling Company. This property was operated for a short time under lease to Carnduff & Duncan, who made a light production.

The Dante Gold Mining Company. The Dante mine on the southwestern slope of Bull Hill has been active under lease to the Big Toad Gold Mining and Milling Company, who are making a steady production. This company has remodeled the Reid Mill and is now treating 50 tons of low grade ore per day. The main shaft has been sunk 32 feet to a depth of 950 feet and 500 feet of crosscutting and drifting has been done.

The Victor Mine, owned by the Moffat Estate, and located on Bull Cliffs, is being operated by the Komat Mining Company, who hold a five-year lease on the property. The company has made a light production and has done some development work.

The Phoenix Deep Mining Company is operating the properties of the Anchoria-Leland Gold Mining Company and Ethel Louise Gold Mining Company under lease. The Conundrum incline shaft has been sunk to a depth of 1,500 feet, where development is now in progress.

The Dig Gold Mining Company operates the Alpha and Omega on the southern slope of Gold Hill, the shaft has been sunk to a depth of 270 feet during 1918, and a small electric hoisting plant and compressor installed. Development was in progress most of the time during 1918.

The Gold Bond Consolidated Mines Company during 1917 was operated under lease to the Gilliard Tungsten Mining and Leasing Company, who made light production. It is now being operated by Hamilton and Son, who have a very good showing near the surface.

The Midget Consolidated Gold Mining Company operated steadily and has made a fair production. They were compelled to discontinue operations early in 1918, owing to war conditions.

The Jerry Johnson Gold Mining Company. The Jerry Johnson mine was operated by the Cripple Creek Deep Leasing Company in 1917, and the lease surrendered early in 1918 and taken over by F. H. Denman, who is making good progress. Ore has been opened up at a depth of 950 feet, the deepest workings on Ironclad Hill, and at this time a production of about a car per week of a good grade milling ore is being made.

The Pride of Cripple Creek. This mine was operated for several months by E. H. Bee and H. C. Nelson, who did considerable development work and made a light production.

The C. O. D. Mining and Development Company prosecuted work on the C. O. D. mine early in 1917 and lessees shipped a light tonnage of milling ore.

The United Gold Mines Company. This company's holdings are operated under the leasing system. During the years 1917 and 1918 there was approximately 8,650 feet of development work done by the company and lessees, and the approximate gross production was $1,109,892.28. In both 1916 and 1917, a Christmas dividend of one cent a share, amounting to $40,000.00 each year, was declared.

The Trail Group has been the most productive of the company holdings during the past two years. This property adjoins the famous Cresson Mine, and is practically surrounded by producing mines. The lessees completed nearly 2,000 feet of development in the last year.

The Roosevelt Drainage Tunnel entered the Trail property early in the year, at a depth of 2,000 feet, and immediately started driving into their territory from the tunnel level. About 400 feet of drifting has been completed and several veins were cut, which give promise of shipping ore. Some prospecting by diamond drill was also carried on from the tunnel.

The W. P. H. Group is located in the most productive section on Iron Clad Hill.

The W. P. H. claim has been a constant producer from the northward development of the vertical fissures and the successive flat veins which were encountered in the progress of operations since 1898. The shaft is 900 feet deep and a good production has been maintained from the bottom levels.

The property has been operated steadily under lease, but has been inactive since December 1, 1918.

The Deadwood Group is operated through two shafts, numbers one and two, being respectively 800 and 850 feet in depth. The claims have been developed by 8,800 feet of drifts and crosscuts and have a production of nearly $1,500,000 to their credit. The Deadwood Leasing Company is operating the Trachyte claim, and is producing a considerable quantity of ore that averages about one ounce gold per ton.

The ore is being mined about 50 feet from the surface and the ore shoot seems to be leading out into the country where the junction of the Deadwood vein with the veins of the Shurtloff and Findley properties might be encountered.

The Pinnacle Group, comprising about 40 acres on the northeast slope of Bull Hill, adjoining the Isabella property, has been inactive during the past year. The production of the Whipp shaft on the Glenn vein is about $25,000.

The M. K. & T. Group. The Bonanza claim of this group has been producing two-ounce ore for the greater part of the past two years. It is located on Squaw Mountain, and has been under lease to the Granite Gold Mining Company. The ore body has already been cut on the 16th level of the Dillon, where the streak assays about 8 ounces in gold per ton.

The Doctor Jack Pot Mining Company recently retimbered its shaft down to the 4th level and expect to resume operations and active development work shortly. The Gregory aerial tram is being removed to facilitate the handling of dump ores from the Ingram, Elizabeth Cooper, Doctor Jack Pot and Morning Glory Mines. The line of the tram has been changed so that the railroad cars may be loaded at the Mary McKinney Mine switch. In the past year about 20,000 tons of dump ore was transported over this tram and shipped to the Golden Cycle Mill at West Colorado Springs for treatment.

The El Paso Consolidated Gold Mining Company is located on the west slope of Beacon Hill, and has been operated entirely on the leasing system.

Trankle & Company, leasing on the C. K. & N. vein, between levels 8 and 9, are shipping ore valued as high as $50 per ton, at the rate of seven to eight cars per month. Paul Grenert has a good showing of ore between levels 4 and 5 of No. 1 shaft, and maintains regular shipments. During the last two years a total of nearly 3,000 feet of development work has been completed and the gross production is valued at $125,000.

The Buckeye Mining and Milling Company, operating the Blue Bird Mine on Bull Hill, has been making regular shipments · of a good grade of ore. It has been operated by the leasing system until November 1, 1918, when operations were discontinued because of the influenza epidemic.

Considerable prospecting was done on Carbonate and Womack Hills and a light tonnage was mined from the Kitty Wells and the Shoo Fly. On the outlying hills prospectors were working for a few months during the past two years on Trachyte, Rhyolite and Copper Mountains.

Dump production in the Cripple Creek District: Howard dump, 22,810 tons; Gregory dump, 13,587 tons; Wedge dump, 351 tons; El Paso dump, 1,020 tons; Economic dump, 45,229 tons; Stratton estate, 3,533 tons.

CUSTER COUNTY

From all present indications Custer County is likely to jump into prominence and be one of the most favored metal mining counties of the state. This condition is being brought about by the demand for metals found here and the increased price of silver. Custer County will no doubt show a favorable record in 1919.

The Buffalo Hunter Mining, Milling and Development Company. This company has adopted the open pit system of mining for the reason that the ores contain a fair value in silver and manganese from the surface to a depth of twenty feet and there seems to be no limit to the width. The new concentrating mill when completed and equipped with the necessary machinery will have a capacity of about 100 tons per day.

The Lady Franklin Mine, operated by Shaeffer and Eichhorn, has a very good showing at a depth of 130 feet. The ores contain 18% zinc, about 35 ounces silver and $6 gold and with reasonable treatment and transportation charges the mine should be operated at a profit.

The Passiflora Mining & Milling Company. This is the most successful operating company in the Hard Scrabble Mining Distriet. The main shaft has been sunk to a depth of 300 feet. Besides the company's operations, three sets of lessees are operating on the property.

The production from this property is about 400 tons per month and the average value of the ore is about $20 per ton. The number of men employed, including lessees, is about 38.

The Maxine Elliott Mine. This property is operated fairly successfully by Traylor and Donaldson through a shaft about sixty feet deep. The vein is three to four feet wide. A small monthly production of fair grade ore is being made.

The Bull Domingo Developing and Mining Company has started operations on the old Bull Domingo Mine and the expense of repairing the shaft and shaft equipment is likely to be heavy. The company has leased, and is repairing, the Passiflora Mill, and when the known ore bodies are reached it will likely be again in the shipping class.

The Westcliffe Mining and Milling Company has built a new mill, known as the Silver Bar Mill, and is operating the King of the Carbonates Mine. The mill has a capacity of 100 tons per 24 hours and operations are fairly successful.

The Luella Mine. A. B. Ferguson and M. L. Mack have installed a small steam plant on this property, and are operating through a shaft about 200 feet deep with very encouraging prospects.

The Princess Annie Copper Mining and Development Company. The mine is located just above timberline on the northeast slope of Crestone Peak, on the Sangre de Cristo Range, about 18 miles southwest of Westcliffe. The ore carries copper values. In order to get the ore to market it will be necessary to build an aerial tram from the valley to the mine, a distance of about 10 miles.

The Peerless Consolidated Copper Company. The development consists of several thousand feet of tunnel, drifts and cross-cuts. The mine is located about 6 miles from Hillside. The prospects are very encouraging for production in the near future.

PUEBLO COUNTY

The United States Zinc Company: This plant has been operating almost continuously since 1902 and has a capacity of 125 tons per day. The change house is one of the most complete in the state; each employee is provided with a steel locker for his clothes and an individual locker for his lunch bucket; the sanitary conditions are good, as are almost all safety conditions throughout the plant. On account of labor shortage the plant is only running at about one-half its capacity. There are about 250 men employed.

The American Smelting and Refining Company. This smelter has been in operation about 41 years, and has treated 1,000 tons per day. In the month of May, 1918, the company had an average of 243 men, and in order to maintain that average it was necessary to hire 666 men. This condition was mostly due to the inferior class of labor caused by war conditions.

The Turkey Creek Stone, Clay and Gypsum Company, operating at Stone City, about 25 miles from Pueblo, is operating three clay mines. The clay is hauled in cars by horses to the kilns, which are operated by The Pueblo Clay Products Company, a distance of about 1,500 feet. The production of finished product is about 40 tons per day. Twenty-five men are employed in both places.

The Dugal Gillespie Quarry, located about 4 miles from Stone City, is producing about 100 cubic feet of building stone per day, and as the stone-dressing plant at Stone City is not in operation, the product is shipped to Denver. Three men are employed.

The Pueblo Lime and Quarrying Company. The quarry is located at Livesey, about 10 miles from Pueblo. The production is about 3,000 tons per month. Seventeen men are employed.

The Colorado Fuel and Iron Company. This company is operating one of the largest quarries in the state at Lime, Colorado, about 9 miles from Pueblo. The operations are conducted systematically, and with great care for the safety of the employees.

The production from the mill is 900 tons per 8 hours and about 300 tons of uncrushed rock per 8 hours. This quarry has been operated most of the time since 1877. There are 102 men employed.

FREMONT COUNTY

Canon City, Colorado

The Empire Zinc Company is running full capacity and treating about 250 tons per day. One hundred men are employed.

The Ohio Zinc Company is a new organization. Their plant cost over $90,000, and is operated at full capacity, working three shifts. The output is 6 tons of zinc oxide per day with a total of 40 employed men.

The Frankenberry Quarries (4), employing 54 men, is making an output of 5,000 tons per month, which is shipped to the Standard Fire Brick Company, Pueblo, Colorado.

The Jewett Quarry is employing five men and making an output of 60 tons per day.

Florence, Colorado

The River Smelting and Refining Company. This plant has a capacity of 2,000 tons per month. They employ 105 men.

Portland, Colorado

The Colorado Portland Cement Company employs 150 men, and is making an output of 2,500 barrels per month.

Concrete, Colorado

The United States Portland Cement Company is making an output of 1,200 barrels per month, and are employing 115 men.

The Oxide Mine is located about midway between Canon City and Texas Creek, on the line of the Denver and Rio Grande Railroad. The property is owned and operated by Messrs. Edingtons, Clark and Isenhour, of Texas Creek, Colorado. The copper occurs near the surface in a vertical vein, as oxide and carbonates, but in depth changes to sulphides.

The Green Mountain Mining Company was operated under bond and lease by John L. Farrell, of Salida, for several years, and made a fair production. The mine was closed down early in the summer of 1918.

EL PASO COUNTY

The Golden Cycle Mining and Reduction Company. The plant of this company is located in West Colorado Springs, and was operated continually during 1917 and 1918. Notwithstanding the fact that the company has operated under adverse conditions during the period of the war and the influenza epidemic, they handled an average of 30,500 tons per month for the two years, employing 170 men, including the office force.

There has been no activity in metalliferous mining in Pueblo, Huerfano, Las Animas and El Paso Counties during this biennial period.

Great credit is due the operators of District No. 2 for their co-operation in installing further safety appliances and enforcing the safety rules of the State Bureau of Mines.

REVIEW OF DISTRICT No. 3

By INSPECTOR R. J. MURRAY

THE LEADVILLE DISTRICT

The Down Town Mines Company has shipped approximately 70,000 tons of ore in the past year, and has been one of Leadville's steady producers. Most of this ore was hoisted through the Penrose shaft, while a fair tonnage, included in the company total, was extracted through sub-leases on the Grey Eagle, Hibschle, Bohn and P. O. S. shafts. This mine produces seven different metals in their various kinds of ore, namely: gold, silver, lead, copper, zinc, manganese and iron.

The Penrose shaft is 875 feet deep, and is the drainage shaft of the Downtown District. During the past year they encountered some very large bodies of silver-rich lead carbonates, while their manganese and silver iron in the upper workings have been a steady source of production, and in pushing ahead their development they have opened some large bodies of ore. Their centrifugal pumps are capable of handling 5,000 gallons per minute, but the steady flow of water for some time has been around 1,450 to 1,500 gallons per minute. This mine is equipped with a large double drum hoist, electrically driven, with cages and all modern appliances. The shaft is $4\frac{1}{2}$x9, 875 feet deep, and is connected with several other mines in the Downtown District. This company retimbered the Coronado shaft within the past few months for the purpose of using it as a hoisting shaft for the new ore bodies discovered in that section.

The Thurn Leasing Co. is developing a fair sized area of ground and producing a fair tonnage of silver-iron ore and also a small tonnage of lead carbonate, with occasional rich streaks of silver chloride.

Harvey, Daniels & Co. have operated the Grey Eagle shaft of the Down Town Mines Co. continuously for the past year and produced a steady tonnage of manganese iron until their ore bodies became exhausted, about last July. They have done considerable prospecting work since.

The Bohn shaft of the Down Town Mines Co. is operated by P. B. Horrigan & Co. This shaft produced a large tonnage of high-grade manganese iron until demand was cut off by the poor market in November. This property has been idle since, but the lessees are at present planning to resume work soon.

The P. O. S. shaft of the Down Town Mines Co., J. J. McInnes & Co. lessees, have been operating continuously for the past 18 months, and have sent out a very small amount of ore. This mine was formerly a very high-grade "silver-iron" producer in the Downtown District.

The Carbonate shaft, Edna Mining Co., began operations in the summer of 1917, and in August of last year struck some large bodies of manganese ore, from which a large tonnage was produced. This ore is said to be the highest grade manganese ore ever shipped out of Leadville in quantities. They are at present working about 18 men, all on development work.

Since the Western Chemical Manufacturing Company purchased the Greenback Mine in May of last year, they have produced a large tonnage of sulphide ores, carrying silver, zinc and gold values, besides iron pyrites for its sulphur content, and a fair tonnage of silver-iron ore. Since this company has taken the property over they have altered and improved all of their surface equipment, particularly their first motion hoist, it being one of two hoisting engines in Leadville that is equipped with the Welch Hoisting engine controller. This mine is also installing automatic stokers on their boilers. Their shaft is 4½x9, 1,335 feet deep, one of the deepest shafts in Leadville.

The Iron-Silver Mining Co., through the Mikado shaft, has been one of Leadville's largest and best producers, having shipped in the past year 35,000 tons of sulphide ores containing silver, lead and zinc values. This shaft is 5x15, 1,130 feet deep and is equipped with one of the best hoisting plants in this district, and is one of the two hoisting engines that is equipped with the Welch hoisting engine controller.

The North Moyer & Tucson shafts of this company, having exhausted their large and productive ore bodies, have ceased operations.

The South Moyer shaft, one of the oldest producers in Leadville, is being worked by Billy Carson & Co., with a force of 20 men, and is sending out a daily tonnage of silver-lead-zinc sulphide ore to the smelter.

During the past year the Iron-Silver Mining Co. started retimbering and enlarging the Pyrennes shaft, and has installed an up-to-date hoisting plant with an 80-foot head-frame, suitable for fast and deep hoisting. The retimbering of this shaft is nearing completion.

The Ibex Mining Company's properties on Breece Hill operated continuously throughout the year and produced a heavy tonnage of gold, silver and copper ores, but their development work was much less than the preceding year, owing to the shortage of labor, which was keenly felt at the Ibex properties. The number of men employed on the property decreased from 250 to 125 during the first half of 1918. The leasing concerns decreased from 50 to 35.

No. 1 shaft was retimbered during the past year, and this took several months. Shipments were curtailed to a certain extent. During the past year this company shipped approximately 30,000 tons of ore of various grades. Some of the extremely rich high-grade pockets were mined in the past year.

The Connors Lease on No. 2 shaft, the Kyle Lease on No. 2 shaft and the Burton Lease on No. 4 shaft were the most successful in the production of high-grade ores.

Development at the properties of the New Monarch Mining Co. has been seriously handicapped during the past year by labor shortage, but indications point to a good year of prospecting activity. They have a proposed drift in view from the Yak Tunnel, which will cut their lower workings 80 feet below the present ones. The New Monarch shaft, the main hoisting shaft of the company, is 760 feet deep at present. They are making preparations to unwater the lower levels which were drowned in the summer of 1917. This company has about 10 set of leasers operating and are working about 10 men on company account, and have been shipping 800 tons per month of oxide and sulphide ores, but at present are storing their ores, awaiting renewal of their contracts.

A shortage of labor that continued throughout the year halted the proposed plans for development work at the Louisville Mine, owned by E. A. Hanifen and W. O. Reynolds, and finally resulted in a complete shut-down last November. Diamond drill holes put down to a considerable depth disclosed a continuation of their ore bodies as large and as rich as those in their upper workings. It is expected that operations on this mine will be resumed very shortly.

The C. J. Dold Mining Company has made steady progress during the past year and has been one of Leadville's best manganiferous iron producers. This company reopened the old Northern shaft, and has shipped 15,000 tons of manganese and silver-lead ore during 1918. They also sunk the great Northern to the second contact, and have opened up extensive bodies of iron, both silver and manganese, also some lead carbonates. They have also retimbered the Newell shaft and have installed a modern plant of hoisting machinery and cages. This will probably be the main hoisting shaft for this property, as it is connected with the Downtown switch on the Denver & Rio Grande Railroad.

The Doty Manganese Company, organized by Mrs. Florence Doty, began operations on the McDonald shaft in Poverty Flats, and sunk it down to the first contact, 275 feet. When the slump in the manganese market came, the property was closed down temporarily.

The A. V. Leasing Co. working at the A. V. shaft was reorganized in March of last year (by local business and mining men of Leadville). A great deal of new development work was done and bodies of manganese iron running 38 to 40 per cent. manganese were opened. This company was fortunate in having their contracts run until July of this year. This lease is employing about 20 men and is keeping up a good tonnage.

The Leadville District Mining and Milling Co. (Mr. Clarence Jarbeau, lessee), operating the Home Extension shaft, has closed down temporarily owing to the slump in the manganese market. They were working 5 men and shipping about 300 tons of manganese iron ore per month.

The Jason shaft in Poverty Flats, worked by G. P. Goodier and Co., and shipping about 25 tons daily of manganese ore, closed down the latter part of December.

Cramer & Co., working on the Star No. 5, Ladder, Waterloo and Yankee Doodle shafts, on Carbonate Hill, have employed a large force of men for the past year, and were shipping a large tonnage of manganese iron ore. At present they are working but 15 men and sending out 25 tons daily of silver-iron ore to the plant of the American Smelting & Refining Company at Pueblo.

Buchanan & Co. are operating the Porter shaft with a force of 12 men, and shipping 400 tons per month of silver-iron ore to the A. S. & R. plants at Pueblo and Leadville. Until the slump in the manganese market they were heavy producers of manganese iron ore.

M. A. Nicholson has operated the Big Chief and Castle View claims through the Castle View shaft for the past year with a small force of men, and has shipped a lot of zinc ore.

The Western Mining Co., operating the Wolftone shaft and other properties, employed a good force of men and produced a good output of zinc carbonate and sulphide ores. They ceased mining operations in November, but their pumps are still running because of an agreement with the Iron Silver Mining Co. and the Greenback Mining Co.

The Yak Mines Co. had a fairly good year, and employed an average force of from 125 to 150 men, but were handicapped by a labor shortage during the latter half of the year. They did not push their development work to any great extent, but have started a campaign of development and exploitation recently and have reopened the Dolly B. workings, which had been abandoned, and retimbered their tunnel wherever it needed it. They have devised a new scheme on their trolley wire. In case of a cave in the tunnel breaking it, they can run to their last sectional switch nearest while the wire is being repaired.

This company has shipped in the past year approximately 57,000 tons of ores, as follows: 34,000 tons iron-silver sulphides, 21,837 tons of zinc-lead ores to Blende, Colo., to the Western Chemical at Denver, to the Sand Springs, Okla., and to Coffeyville, Kan., and 1,800 tons of pyritic ores to various acid makers.

The lower levels of the Cord and Whitecap winzes were flooded by water in May of last year. This contained an unusual amount of sulphuric acid, which corroded their pumping apparatus so that it was useless. After installing a new pumping plant and finding a means of neutralizing the acid in the water, it took several months of hard pumping to recover the lower levels of the two winzes which had been very productive at the time they were drowned.

The Derry Ranch Dredge continued to work throughout the year and produced close to $100,000 in gold.

PARK COUNTY ·

There was considerable activity in Park County during the past two years. Several of the old silver producers are being opened and a material increase in production is expected if the price of silver remains near a dollar an ounce.

The Louisiana-Colorado Mining Company completed a new 250-ton mill and are preparing to start operations on a large scale.

The Whale Mines Leasing & Reduction Company completed a 75-ton mill and has shipped about 100 tons of silver-lead concentrate.

The Hock Hocking shipped some high-grade lead-silver ore.

The London Mining & Milling Company drove their new tunnel to within 400 feet of the vein. It is now 1,600 feet long.

The Mill Gulch Mining Company continued their development work on a body of copper ore.

The Fannie Barrett Mining Company built a new tram in 1918, and began development work on this old silver producer.

The Carbonate King Mining Company, at Guffey, and the 4 J Mining Company, at Alma, ceased operation in 1917.

SUMMIT COUNTY

The Iron Mask Mine on the northeast slope of Shock Hill is sinking a new two-compartment shaft, with three shifts of miners. This mine is equipped with an electric hoist and pump.

The Royal Tiger Mining Company, operating the old 1. X. L. and other adjoining properties with a working force of fifty men, has remodeled its thirty-ton wet concentration, and flotation test mill. Work has also been done on a new shaft sunk below their tunnel level and drifts driven to block out their ore bodies, which consist chiefly of bismuth and gold ores. Within the past year this company has erected saw-mills, planing mills and a number of cottages, boarding houses and bunk houses. Their timber supply is taken from their own properties and put into the finished product at their mills.

The Josie Claim, operated by Paul Adams, has shipped a very small tonnage of high-grade silver ore.

The Carbonate Mine on Mt. Baldy, owned by George Moon, of Breckenridge, has worked a few men and produced a small tonnage of silver-lead ore.

The Quandary Queen Mining Company on Mt. Helen has produced a small quantity of silver-lead ores.

The Brooks Snider, under lease to Ed Auge, has encountered a small streak of high-grade gold ore.

The Jessie Mine, employing 12 men, has ceased operations in their tunnel levels and are at present engaged in sinking a shaft 120 feet below their tunnel level. At this point they intend to do considerable development work. The underground shaft is equipped with an electric hoist and pump.

The Molly B. Mining and Milling Company worked four men during the past season, and produced a small tonnage of bismuth ore with gold, silver, copper and lead values.

Midwest Mining and Holding Company, with four men, is at present engaged in sinking a shaft 160 feet to cut their known ore bodies, encountered in their tunnel.

The King Mine, located on Tucker Mountain, operated with a force of four men.

The Mutual Co-operative Mining Company, operating through the Golden Queen Tunnel and employing about 15 men, is shipping 100 tons per month of an oxidized ore carrying high silver values.

The Silver Edge Mining Company worked during the past year with a force of nine men, and shipped a small tonnage of sulphide ore for its gold, silver and iron contents.

The Elk Mountain Mines Company, employing about 30 men, shipped about 130 tons per month concentrates from their mill, which is 100 tons capacity and wet concentration.

The American Metals Company at Climax was one of the largest producers of molybdenite, and while in operation worked a force of about 150 men and milled 456 tons of crude ore per day. It closed down in March to await a change in the market. This mine and mill is thoroughly equipped with up-to-date mining and milling machinery. It maintains its own hospital in conjunction with first aid class and mine rescue teams. This company has erected during the past year well-heated and ventilated bunk and boarding houses for the men and also dwelling houses for families.

The Wellington Mines Company, employing on an average of about 125 men, produced a very large tonnage of high-grade zinc ores containing silver-lead values. During the last few years the Wellington Mines have been one of the largest producers of zinc ores in the state, but owing to the depression in the metal market, this company has suspended shipping operations, diverting its attention to new prospect work and also developing and blocking out its large body of high-grade zinc ore. This mine is thoroughly equipped with machine shop, saw-mills, etc. Within the past year they have erected at the mine a new and modern boarding house, finished with hardwood throughout.

EAGLE COUNTY

Empire Zinc Company has employed an average of 225 men, mines and mill, inclusive. They are mining 6,000 tons per month, 5,400 tons of which goes to their mill at Belden Switch for concentration, the remaining 600 tons of crude ore to the A. V. plant at Leadville, the Ohio and Colorado at Salida, and the Western Chemical at Denver. Their concentrates are shipped to their plant at Canon City. This company has started an innovation in the way of prospecting in the use of a Dobbins Core Drill. The cutting surface of the Dobbins Core Drill bit is chilled shot, automatically poured into the casing of the drill. A small amount of building has been done on Astor Flats during the past year, where the mill will be located eventually, the intention being to move it from its present location in Eagle River Canon. Most of their work is being done through Nos. 1 and 2 inclines, No. 1 being 1,800 feet, No. 2 being 1,400 feet. The ores are hauled to the foot of these inclines by mules.

The T. P. M. Leasing Company, operating on the Poorman and Tram properties, and employing four men, are shipping 30 tons of sulphide ore to the A. V. plant at Leadville.

The Mable Mine, with several sets of lessees, has been shipping a small tonnage of ore to the A. V. plant at Leadville for its gold, silver and copper contents.

The Wyoming Valley Placer Group has worked with a force of four to six men as lessees, and has produced a small tonnage of silver-lead ore.

The Bowie Mine has worked three men during the past summer.

The Wheat Mine worked two men for a few months during the past year.

The California did a small amount of development work during the last year.

The Little Chicago Group operated with a few men during the past biennial period.

CHAFFEE COUNTY

, The Hidden Treasure Mining and Leasing Company, operating almost continuously during the past two years, has kept up a steady production of a good grade of copper sulphide ores, which were shipped to the A. V. plant at Leadville. A small tonnage of 'zinc ores has been shipped to the various zinc smelters. Their ores are being carried from the mine to the railroad switch by a track tram operated by a hoist, the length of which tram is 9,000 feet. The employees of the company average about 20 men.

The Mary Murphy Gold Mining Company, with a force of about 125 men, has operated continuously. The greater part of its tonnage, aggregating about 250 tons per month, is being shipped to the A. V. plant at Leadville for its gold-silver-copper-lead contents. A small additional tonnage is being shipped to Henrietta, Oklahoma, for zinc, and to the Ohio and Colorado at Salida for silver-iron. The 200-ton wet concentration, flotation and electro-static separation mill has been working intermittently on their old tailing dump for the past year.

The Monarch-Madonna Mining Company, operating on the upper tunnel levels of the Madonna Mine, has produced a considerable tonnage of carbonate of zinc ores and a small tonnage of silver-lead-carbonate ores. The company ceased active operations on No. 6 Tunnel during September of last year. The entire property is being worked by lessees.

The Giant Eclipse Mining Company has been working the lower levels of the Eclipse with a small force of men. The remainder of the property is being worked by lessees. Two cable trams, 1,200 and 1,800 feet, respectively, have been erected to carry ore to the railroad switch. With an average force of seven men, this property produced an average monthly tonnage of 150 tons of carbonate of zinc.

The Turret Copper Mining and Milling Company at Turret is working a few men on development work.

The Ohio and Colorado Quarry at Garfield is working eight men taking out lime rock.

Fred L. Watson has one man on the Christmas, and is operating through Lily Tunnel.

The Garfield Mining Company operated throughout the year, and with a force of about eight men, put out a monthly tonnage of 100 tons of a sulphide ore carrying gold, silver and iron.

The Allie Belle Mine has shipped some very high-grade ore and is employing at the present time five men on development work. They could produce a good tonnage were they not handicapped by a lack of railroad cars.

The Flora Belle Mining Company operated almost continuously with a small force of men during the past year.

The Independence was worked with two men during the summer season.

The Monarch Tunnel and Development Company worked three men during the past year on development work.

PITKIN COUNTY

The Smuggler Leasing Company, with Mr. Anderson in charge as manager, has carried on extensive operations on their property during the past two years, working through the Molly, Free Silver and Smuggler shafts. During the month of December, 1918, this company pulled their pumps, thereby abandoning their lower workings on Smuggler Mountain. Until the abandonment of these workings, these shafts were producing a fairly large tonnage of crude and milling ores, which were shipped to the A. S. & R. plants at Leadville, Pueblo and Denver. While the pumps were in operation, 2,400 gallons of water per minute at a 1,200-foot head were being pumped through the Free Silver shaft.

At the present time the work of this company is being done on No. 2 Tunnel and Cowenhoven Tunnel on Smuggler Mountain, and on the Durant and Ajax Tunnels on Aspen Mountain. During the past two years No. 2 Tunnel has produced an average shipment of 80 tons daily of lead-silver ores, same going to Smuggler Mill for concentration.

The Cowenhoven Tunnel, with 12 men, company account, and three sets of lessees, has been shipping about 120 tons per month to the A. S. & R. plants.

The Ajax and Durant Tunnels, employing approximately 135 men, lessees and company men, inclusive, have produced a daily tonnage of from 40 to 65 tons. This ore is also being shipped to the A. S. & R. plants. At present the Durant Tunnel is being extended to cut the West Aspen fault.

The Newman Mining, Milling and Leasing Company, after cleaning out their old mine workings, has, under the present management, become one of Aspen's best silver and lead producers. Since reopening, this company has employed an average of about

25 men and during the past few months has been producing about 25 tons of crude ore, which is being shipped to the Smuggler Mill at Aspen for concentration. During the past year this company has averaged in shipments to the A. S. & R. plants about 750 tons per month of silver-lead-carbonate ore.

During the past year work was energetically pushed on the Park Tunnel. Power houses, bunk houses, etc., have been erected, and since the installation of machine drills this tunnel has made rapid progress, its length now being about 800 feet. At the present time this tunnel is being driven to reach some of the old silver-lead producers of the once-famous Tourtelotte Park District. If successful, this tunnel will strike these old properties at a depth considerably lower than any of the old workings.

The Hope Mining, Milling and Leasing Company, working on the Hope Tunnel, has reached a total length of nearly 7,000 feet, and during the past year has encountered several small stringers of ore. The average number of men being employed by this company is five, and the work is being done almost entirely with machine drills.

The Spar Consolidated Mining Company, operating through the Veteran Tunnel with several sets of lessees, has been making a steady shipment of about 40 tons per month of silver-lead ores to the Aspen Sampler at Aspen.

During the past year the Aspen Contact Mining Company, now under lease to the Smuggler Leasing Company, has begun operations on this property with a force of 12 men. At the present time this company is engaged in sinking a shaft whose depth is 64 feet. Small bunches of ore have been encountered. This mine is located seven miles from Aspen, and in a practically new mineral belt.

The Hurricane Mining and Milling Company has reopened the Climax and Etc. Tunnels after an idleness of 25 years' duration, and has done considerable prospecting and development work.

The Pride of Aspen has worked a small force of men during the past biennial period.

The Midnight Mining Company did a small amount of development work during the past year with a force of five men.

REVIEW OF DISTRICT No. 4

By INSPECTOR ROBERT INNES

District No. 4 comprises 15 counties in southwestern Colorado. It is known and designated as the San Juan District. Within the boundaries of this big mining area is found a network of veins running at all points of the compass and containing almost every known mineral. It is said without contradiction to be the strongest mineralized section in Colorado. The biennial period ending November 30, 1918, witnessed considerable activity in this district, despite war conditions. Scarcity of labor, high cost of explosives, shortage of chemicals, increase in freight and smelting rates operated to check production from mines in Colorado.

The records will show that the number of men employed and the production from the mines in this district will compare favorably with the previous biennial period, when conditions were more favorable.

' Some of the big companies maintained employment offices in Denver, and had no difficulty in securing sufficient men to operate their properties. The chief difficulty was in securing competent men, and, were it not for the shortage of experienced miners, this district would have given a better account of itself during the period under review. The most notable feature of this period was the advent in the San Juan mining fields of the Tonopah Belmont Development Company, who secured control of the Alta Mine near Telluride, and the United States Smelting and Refining Company, who secured control of the Sunnyside Mines near Silverton. Both of these companies have constructed large concentration plants on their properties and have made other improvements looking toward an increased production from their mines. In the early days of the district, milling processes were crude. It has been truly said that no better than 60% saving was made, but with the installation of modern processes, such as flotation, electro-magnetic separation, etc., recoveries better than 90% are made, as will be seen by a visit to the mills of the district using these processes. Space will not allow a detailed account of each individual operation, but I will endeavor to give a general review in as few words as possible.

The upper San Miguel District, where Telluride's big four mines are located, namely, the Liberty Bell, Tomboy, Smuggler Union and Colorado Superior, has yielded the greater part of San Miguel County's production.

The Belmont Wagner Mining Company completed and put into commission their new mill in July, 1918, but were compelled to cease production early in the fall owing to influenza conditions. In the lower district the Primos Chemical Company, operating mines at Bear Creek, Fall Creek and Leopard Creek, employed a large force of men and produced a large tonnage of vanadium ores.

The Carbonero, Caribou and Favorite Mines were the principal operators in the Ophir District, and shipped considerable ore. The Wagner Brothers secured a lease on the Cimarron Mine, and shipped some dump ore. The Caruthers Mill, operating on ore from the Sheridan crosscut, was kept busy. The mill at the Lewis Mine was equipped with flotation machinery and operated to its full capacity, 25 tons daily. Some development work was done in the Bear Creek region. In the Silverton District the Sunnyside Mining & Milling Company discontinued production during the main part of 1917 and 1918, while a large force of men were employed on construction and development work. The new 500-ton mill which is equipped with the latest machinery, including selective flotation, was put into commission May, 1918, and operated continuously until they were compelled to close down in November, owing to a shortage of men caused by the influenza epidemic.

· The Kittimac closed down in the winter of 1917. The Hamlet operated until the fall of 1918. The Shenandoah, Highland Mary and Pride of the West were steady producers until the fall of 1918, when the two last named ceased operations owing to the drop in the price of base metals. In Arastra Gulch and Basin the Silver Lake, Iowa and May Flower Mines have been steady producers, some of the ore being shipped to the smelters and the balance treated in the Iowa Mill. In the outskirts of Silverton, the Lackawanna, Coming Wonder, North Star, Dora, Champion and Detroit-Colorado Mines produced considerable ore and did a great deal of development work, and the North Star Mill operated as a custom plant. In the Gladstone District the Gold King Leasing Company discontinued operations in 1918 on the Gold King properties, at which time the properties were taken over by the Gold King Extension Mines Company, who own an adjoining property. The Henrietta Copper Mining Company was incorporated in 1917 to operate the Henrietta Mine in Prospect Basin, which has been inactive for several years. The Joe and John Mine, also in Prospect Basin, was active and shipped some good ore. In the Animas Forks section the Columbus and Hermes properties were operated on development work. The Congress, St. Paul and San Antonio Mines at Red Mountain were active and shipped considerable ore.

The main feature in Ouray County during the biennial period was the completion of the Camp Bird Tunnel. The Camp Bird vein was reached at a distance of 11,000 feet from the portal of the tunnel, and development work has since been conducted, with three shifts of machines driving along the vein.

The Atlas Mine has equalled its production of former years.

In Virginia Basin the Mountain Top Mine operated continuously. An underground concentrating mill was completed and put into commission in 1917, and an aerial tramway constructed to a point down the gulch, safe from snow slides.

The Calliope, Bachelor-Khedive, Wedge and El Mahdi Mines were operated by lessees and shipped considerable ore. In the Red Mountain section a strike of rich gold-silver ore in the Guadaloupe and the discovery of fluorspar in the Barstow created some activity. The Joker Tunnel, Yankee Girl, Genesee Vanderbilt, White Cloud and other smaller properties shipped some ore, but worked mostly on development work.

The Yellow Jacket, the Uncle Sam and the Silver Link were active on development work.

Part of the Ouray Smelter building was equipped with the latest milling processes, including flotation, and put into commission in June, 1918, to handle custom work.

In the Rico District operations were curtailed by a shortage of men in 1918. The production from the Rico Wellington Mines was not as large as during the previous biennial period, but a great amount of development work was done in the mines. In the Silver Creek section the Rico Consolidated, Rico Argentine and Marmatite Mining Companies were active and shipped considerable ore.

The Rico Mines Company on Newman Hill had a few men employed on repair work. The Marmatite Mining Company secured a lease on the Pro Patria and began operations in the fall of 1918. Denver parties sank a shaft on the Swickheimer property on C. H. C. Hill, prospecting for ore beds. The Cadiz Mining Company conducted development work on the Lillie D. Mine and shipped a few cars of ore.

In Horse Gulch the McIntyre Mining Company operated the Puzzle Mine on development work, and the Calico Peak property in the fall of 1918 was put in shape to conduct development work during the winter. There was not much activity in the La Platas. The May Day was inactive. The Valley View (Idaho Mine) was operated under lease with about four men employed. The Jumbo was operated by lessees with a force of four men, and shipped some ore. The Allard Mining Company, The Lewis Mountain Mining Company and The Cumberland Mines Company each employed a small force on development work.

In the Lake City District, The Highland Chief, Ocean Wave, Ute and Ulay, Hidden Treasure and the Black Crook Contention properties were active, and besides doing considerable development work, they shipped some ore.

In the White Pine Section of Gunnison County, The Akron Mines Company operated their mill part of 1917 on ores from their various properties. In 1918 they constructed a new mill at the mouth of the Akron Tunnel, and in the meantime work was almost discontinued in the mines. Two men were employed continuously by The Gunnison Copper Company driving a tunnel at Tomichi. The discovery of graphite and molybdenite near Pitkin created considerable activity in that section. The Bon Ton Mine has been operated by the Pennsylvania Molybdenum Mines Company for molybdenite, and the Ethiopian and Avoca Virginia properties were operated for graphite.

The Camp Bird Mining, Leasing and Power Company, in the Bowerman Section, operated continuously on development work.

The Doctor Mine in the Tin Cup Section was active and produced considerable carbonate of zinc ore. The Good Hope and Headlight Mines at Vulcan and Spencer have been active on development work. The Anaconda Mine, in the same section, was active in 1917 and produced considerable copper ore, but closed down in 1918.

In the Gold Creek District the Belzora Bassick in 1917 had a force employed on development work, and made a trial run of ore in the Santa Cruz Mill, but discontinued operations in 1918. The Carter Mine employed a force of men in 1918 and shipped some ore.

Creede was more active in the biennial period just passed than it has been for a number of years. The discovery of large deposits of fluorspar near Wagon Wheel Gap was the means of employing a large force of men continuously. In the Sunnyside District a rich strike of high-grade silver ore was made in 1917 on the Quintet property. Mining has been conducted in this district since the early days of Creede and considerable ore shipped, but never before has an ore body of such magnitude and richness been discovered. On the Lefevre property, adjoining the Quintet, a similar ore shoot was opened up in 1918. The combining of all the properties along the Bacheler vein under a 20 years' lease to the Creede Exploration Company, a subsidiary of the American Smelting & Refining Company, was effected in 1917, who, in addition to extensive development work, shipped some ore.

The Equity Creede Mining Company, located on West Willow Creek, operated their property almost continuously and shipped considerable silver-copper ore. The Mineral County Mining & Milling Company, who have been operating under lease for several years the properties of the Creede Mining & Milling Company and the Humphreys Mill, practically discontinued operations in 1918. In the upper workings of the Bachelor, Commodore and New York-Chance Mines several sets of lessees have been operating and shipped considerable ore. The Silver King was operated chiefly on development work.

In the Bonanza District of Saguache County, in 1917, the St. Louis, Josephine, Memphis, Cocomongo, Bonanza, Rawley, Shawmut and Eagle Mines were all active and shipped considerable ore, but in 1918 the Cocomongo and Eagle were the only active mines in the district.

In the Paradox District of Montrose County, The Standard Chemical Company and other smaller properties have been constant producers of carnotite ores.

Only prospecting and assessment work has been done in Archuleta, Costilla, Conejos, Montezuma and Rio Grande counties.

In conclusion, it has been my endeavor to cover this district during the past biennial period as often as the limited allowance provided for traveling expenses, would permit. Some of the large properties were inspected two or three times in each year, while some of the smaller properties in outlying districts were only inspected once in each year. I made 177 verbal recommendations and issued 46 written recommendations for safety, which were generally complied with. The co-operation of those in charge of the various properties with this department, undoubtedly has been the means of preventing a number of accidents. If the same spirit of co-operation existed between the employees and their employers, the number of accidents would be reduced considerably.

I wish to express my thanks to those in charge of the various properties at the time of my inspections, for the assistance rendered.

STORAGE AND USE OF EXPLOSIVES

All persons engaged in the metal mining industry in Colorado have adopted careless practices in the handling and use of high explosives, which are not only very dangerous, but have resulted in much waste of explosives. Storage places have been selected for convenience and little consideration given to security of storage. Notwithstanding these conditions, the industry as a whole has shown an unusually good record in the use of explosives; very few accidental explosions of powder in storage have occurred, and there have been few casualties.

The almost universal lack of proper regulations in the entire country for the storage and distribution of high explosives was not realized until the existing dangers were called to our attention by the War Department's recommendation to Congress for the enactment of a suitable regulatory law to control the manufacture, sale, distribution and use of explosives, and the quick passage of such a law (effective November 15, 1917), placing the administrative authority in the hands of the Director of the Federal Bureau of Mines. The provisions of this law, and the rules and regulations adopted by authority of it, are perhaps more drastic than are needed in peace times. The principles and practices established are considered of benefit, generally, and should be enforced through state authority, should the Federal law become inoperative. Therefore, it is recommended that the Commissioner of Mines, under the authority granted him by the statutes of this state, should adopt and publish rules and regulations designed to conform to the general provisions of the Federal law and the regulatory measures which have been adopted by authority of it. For the information of those who may be interested, the Federal law is reprinted in another part of this report.

[PUBLIC—No. 68—65TH CONGRESS.]
[H. R. 3932.]

An Act to prohibit the manufacture, distribution, storage, use, and possession in time of war of explosives, providing regulations for the safe manufacture, distribution, storage, use, and possession of the same, and for other purposes.

Be it enacted by the Senate and House of Representatives of the United States of America in Congress assembled, That when the United States is at war it shall be unlawful to manufacture, distribute, store, use, or possess powder, explosives, blasting supplies, or ingredients thereof in such manner as to be detrimental to the public safety, except as in this Act provided.

SEC. 2. That the words "explosive" and "explosives" when used herein shall mean gunpowders, powders used for blasting, all forms of high explosives, blasting materials, fuses, detonators, and other detonating agents, smokeless powders, and any chemical compound or mechanical mixture that contains any oxidizing and combustible units, or other ingredients, in such proportions, quantities or packing that ignition by fire, by friction, by concussion, by percussion or by detonation of, or any part of the compound or mixture may cause such a sudden .generation of highly heated gases that the resultant gaseous pressures are capable of producing destructive effects on contiguous objects, or of destroying life or limb, but shall not include small arms or shotgun cartridges: *Provided,* That nothing herein contained shall be construed to prevent the manufacture, under the authority of the Government, of explosives for, their sale to or their possession by, the military or naval service of the United States of America.

SEC. 3. That the word "ingredients" when used herein shall mean the materials and substances capable by combination of producing one or more of the explosives mentioned in section one hereof.

SEC. 4. That the word "person," when used herein, shall include States, Territories, the District of Columbia, Alaska, and other dependencies of the United States and municipal subdivisions thereof, individual citizens, firms, associations, societies and corporations of the United States and of other countries at peace with the United States.

SEC. 5. That from and after forty days after the passage and approval of this Act no person shall have in his possession or purchase, accept, receive, sell, give, barter or otherwise dispose of or procure explosives, or ingredients, except as provided in this Act: *Provided,* That the purchase or possession of said ingredients when purchased or held in small quantities and not used or intended to be used in the manufacture of explosives are not subject to the provisions of this Act: *Provided further,* That the superintendent, foreman, or other duly authorized employee, at a mine, quarry, or other work may, when licensed so to do, sell or issue to any workman under him, such an amount of explosives, or ingredients, as may be required by that workman in the performance of his duties, and the workman may purchase or accept the explosives, or ingredients, so sold or issued, but the person so selling or issuing same shall see that any unused explosives, or ingredients, are returned, and that no explosives, or ingredients, are taken by the workman to any point not necessary to the carrying on of his duties.

SEC. 6. That nothing contained herein shall apply to explosives or ingredients while being transported upon vessels or railroad cars in conformity with statutory law or Interstate Commerce Commission rules.

SEC. 7. That from and after forty days after the passage of this Act no person shall manufacture explosives unless licensed so to do, as hereinafter provided.

SEC. 8. That any licensee or applicant for license hereunder shall furnish such information regarding himself and his business, so far as such business relates to or is connected with explosives or ingredients at such time and in such manner as the Director of the Bureau of Mines, or his authorized representative, may request, excepting that those who have been or are at the time of the passage of this Act regularly engaged in the manufacture of explosives shall not be compelled to disclose secret processes, costs, or other data unrelated to the distribution of explosives.

SEC. 9. That from and after forty days after the passage and approval of this Act every person authorized to sell, issue, or dispose of explosives shall keep a complete itemized and accurate record, showing each person to whom explosives are sold, given, bartered, or to whom or how otherwise disposed of, and the quantity and kind of explosives, and the date of each such sale, gift, barter, or other disposition, and this record shall be sworn to and furnished to the Director of the Bureau of Mines or his authorized representatives, whenever requested.

SEC. 10. That the Director of the Bureau of Mines is hereby authorized to issue licenses as follows:

(a) Manufacturer's license, authorizing the manufacture, possession, and sale of explosives and ingredients.

(b) Vendor's license, authorizing the purchase, possession, and sale of explosives or ingredients.

(c) Purchaser's license, authorizing the purchase and possession of explosives and ingredients.

(d) Foreman's license, authorizing the purchase and possession of explosives and ingredients, and the sale and issuance of explosives and ingredients to workmen under the proviso to section five above.

(e) Exporter's license, authorizing the licensee to export explosives, but no such license shall authorize exportation in violation of any proclamation of the President issued under any Act of Congress.

(f) Importer's license, authorizing the license to import explosives.

(g) Analyst's, educator's, inventor's, and investigator's licenses authorizing the purchase, manufacture, possession, testing, and disposal of explosives and ingredients.

SEC. 11. That the Director of the Bureau of Mines shall issue licenses upon application duly made, but only to citizens of the United States of America, and to the subjects or citizens of nations that are at peace with them, and to corporations, firms, and associations thereof, and he may, in his discretion, refuse to issue a license when he has reason to believe, from facts of which he has knowledge or reliable information, that the applicant is disloyal or hostile to the United States of America, or that, if the applicant is a firm, association, society, or corporation, its controlling stockholders or members are disloyal or hostile to the United States of America. The director may, when he has reason to believe on like grounds that any licensee is so disloyal or hostile, revoke any license issued to him. Any applicant to whom a license is refused or any licensee whose license is revoked by the said director may, at any time within thirty days after notification of the rejection of his application or revocation of his license, apply for such license or the cancellation of such revocation to the Council of National Defense, which shall make its order upon the director either to grant or to withhold the license.

SEC. 12. That any person desiring to manufacture, sell, export, import, store, or purchase explosives or ingredients, or to keep explosives or ingredients in his possession, shall make application for a license, which application shall state, under oath, the name of the applicant; the place of birth; whether native born or naturalized citizen of the United States of America; if a naturalized citizen, the date and place of naturalization, business in which engaged; the amount and kind of explosives or ingredients which during the past six months have been purchased, disposed of, or used by him; the amount and kind of explosives or ingredients now on hand; whether sales, if any, have been made to jobbers, wholesalers, retailers, or consumers; the kind of license to be issued, and the kind and amount of explosives or ingredients to be authorized by the license; and such further information as the Director of the Bureau of Mines may, by rule, from time to time require.

Applications for vendor's, purchaser's, or foreman's licenses shall be made to such officers of the State, Territory, or dependency having jurisdiction in the district within which the explosives or ingredients are to be sold or used, and having the power to administer oaths as may be designated by the Director of the Bureau of Mines, who shall issue the same in the name of such director. Such officers shall be entitled to receive from the applicant a fee of 25 cents for each license issued. They shall keep an accurate record of all licenses issued in manner and form to be prescribed by the Director of the Bureau of Mines, to whom they shall make reports from time to time as may be by rule issued by the director required. The necessary blanks and blank records shall be furnished to such officers by the said director. Licensing officers shall be subject to removal for cause by the Director of the Bureau of Mines, and all licenses issued by them shall be subject to revocation by the director as provided in section eleven.

. SEC. 13. That the President, by and with the advice and consent of the Senate, may appoint in each State and in Alaska an explosives inspector, whose duty it shall be, under the direction of the Director of the Bureau of Mines, to see that this Act is faithfully executed and observed. Each such inspector shall receive a salary of $2,400 per annum. He may at any time be detailed for service by said director in the District of Columbia or in any State, Territory, or dependency of the United States. All additional employees required in carrying out the provisions of this Act shall be appointed by the Director of the Bureau of Mines, subject to the approval of the Secretary of the Interior.

SEC. 14. That it shall be unlawful for any person to represent himself as having a license issued under this Act, when he has not such a license, or as having a license different in form or in conditions from the one which he in fact has, or without proper authority make, cause to be made, issue or exhibit anything purporting or pretending to be such license, or intended to mislead any person into believing it is such a license, or to refuse to exhibit his license to any peace officer, Federal or State, or representative of the Bureau of Mines.

SEC. 15. That no inspector or other employee of the Bureau of Mines shall divulge any information obtained in the course of his duties under this Act regarding the business of any licensee, or applicant for license, without authority from the applicant for license or from the Director of the Bureau of Mines.

SEC. 16. That every person authorized under this Act to manufacture or store explosives or ingredients shall clearly mark and define the premises on which his plant or magazine may be and shall conspicuously display thereon the words "Explosives—Keep Off."

SEC. 17. That no person, without the consent of the owner or his authorized agents, except peace officers, the Director of the Bureau of Mines and persons designated by him in writing, shall be in or upon any plant or premises on which explosives are manufactured or stored, or be in or upon any magazine premises on which explosives are stored; nor shall any person discharge any firearms or throw or place any explosives or inflammable bombs at, on, or against any such plant or magazine premises, or cause the same to be done.

SEC. 18. That the Director of the Bureau of Mines is hereby authorized to make rules and regulations for carrying into effect this Act, subject to the approval of the Secretary of the Interior.

SEC. 19. That any person violating any of the provisions of this Act, or any rules or regulations made thereunder, shall be guilty of a misdemeanor and shall be punished by a fine of not more than $5,000 or by imprisonment not more than one year, or by both such fine and imprisonment.

SEC. 20. That the Director of the Bureau of Mines is hereby authorized to investigate all explosions and fires which may occur in mines, quarries, factories, warehouses, magazines, houses, cars, boats, conveyances, and all places in which explosives or the ingredients thereof are manufactured, transported, stored, or used, and shall, in his discretion, report his findings, in such manner as he may deem fit, to the proper Federal or State authorities, to the end that if such explosion has been brought about by a willful act the person or persons causing such act may be proceeded against and brought to justice; or, if said explosion has been brought about by accidental means, that precautions may be taken to prevent similar accidents from occurring. In the prosecution of such investigations the employees of the Bureau of Mines are hereby granted the authority to enter the premises where such explosion or fire has occurred, to examine plans, books, and papers, to administer oaths to, and to examine all witnesses and persons concerned, without let or hindrance on the part of the owner, lessee, operator, or agent thereof.

SEC. 21. That the Director of the Bureau of Mines, with the approval of the President, is hereby authorized to utilize such agents, agencies, and all officers of the United States and of the several States, Territories, dependencies, and municipalities thereof, and the District of Columbia, in the execution of this Act, and all agents, agencies, and all officers of the United States and of the several States and Territories, dependencies, and municipalities thereof, and the District of Columbia, shall hereby have full authority for all acts done by them in the execution of this Act when acting by the direction of the Bureau of Mines.

MILLING OF COMPLEX ORES

As stated elsewhere in this report, many of the mining districts in Colorado contain large tonnages of mixed sulphide ores that appear to be responsive to the newer applications of the preferential principle of the flotation process. The possibilities in this respect have not been generally recognized, hence very little well directed effort has been made to separate these ores into their constituent minerals by the use of this process. Results obtained, however, in laboratory and pilot plants, and the notable success, on a large scale at the Sunnyside Mill, in San Juan County, assure the future of the mining industry in many districts of our state.

OIL FLOTATION

Oil flotation is purely an ore-dressing process. In order to prepare ore for treatment by this process, it is first ground fine enough to free the metallic mineral particles from the "gangue." This grinding is usually done in a limited amount of water and the resultant product is termed a "pulp." When a small quantity of certain oils (usually essential wood oils, coal tar or oils derived from it, or combinations of these) are thoroughly mixed into this pulp, the metallic mineral particles readily adhere to air bubbles which may be formed by introducing air through a porous bottom to the receptacle, or by beating the air into the pulp by violent mechanical agitation. Thus the air bubbles become vehicles which convey such mineral particles to the surface of the pulp and form there a froth which is removed by skimming, or is crowded over the sides of the receptacle by new air bubbles that are constantly rising to the surface and building on themselves. This froth is broken down by collecting it on the surface of a vacuum filter, where the air is removed by spray-washing and settling in tanks. Sufficient moisture is then removed to make the product suitable for shipping loose to the smelters. This product, when properly handled, is as acceptable to the smelters as the concentrates made with gravity machines.

The first plant on this continent to adopt the improved process is said to have been at Britannia, B. C., and the second plant at the property of the Atlas Mining & Milling Company, Sneffels, Ouray County, Colorado. The former started in 1912 and the latter in 1913. Many other plants have been started since that time; in fact, most of the large copper producers of the country doing ore milling are using the process.

Many small mills in Colorado have adopted the process during the past four years, the greater part of this number during the last two years; but, as has been too often the case, the process was heralded as a "cure all," and more failures than successes have been recorded. Some very beneficial results have been obtained, however, in the application of this process to milling molybdenite ores in Summit and Clear Creek Counties. A fair success in treating gold ores has been reported from the Cripple Creek District and there is a rumor, apparently well founded, that a new application of the process has shown unexpectedly good results, which will have a stimulating effect upon the production there.

A method of selective or preferential flotation has been worked out at the Sunnyside Mines of San Juan County, where galena or lead sulphide is floated out of the pulp and subsequently sphalerite or zinc sulphide. It leaves in the tailings an excess of bright iron sulphides which would be a detriment were they contained in either of the selected products. This is apparently the most notable advancement in the application of the process in this state, and promises much toward the solution of the problem of the separation of the complex sulphide ores, of which Colorado has great quantities.

The plant at the property of the Atlas Mining & Milling Company has been in constant operation since it was first installed, but there has never been any effort made to do selective work, for the object was to save all the sulphides and ship them as one product. The lead recovery in this mill averages around 96 per cent and the recovery of silver from one of the silver ores as high as 94 per cent.

There is a good opportunity for application of this process in the hands of skilled manipulators, and the mining industry of Colorado would benefit greatly if a more determined and systematic effort was made to apply this process to the problems of separation and concentration of the complex sulphide ores.

DRAINAGE AND TRANSPORTATION TUNNELS

With the exhaustion of the bonanza ores near the surface, the question of pumping began to be an important one in the mining industry. The early miners were troubled but little by water in their work. The ruggedness of the topography which surrounds most mining districts suggested the use of tunnels for draining the mines. Many small tunnels were started, and we find that Dr. Rossitur Raymond makes the following remarks in speaking of Clear Creek County in 1873:

"On the whole, the tunnel excitement in Clear Creek County has been productive of no good, but, on the contrary, has been of very considerable detriment to the true interests of the county. Aside from the numerous cross-cut tunnels, generally of short length, which have been driven to cut deposits of ore known to exist, there are about twenty tunnels in the county driven across the course of known belts of lodes, and it is a reasonable estimate to say that nearly or quite 10,000 feet of tunnels have been driven, at an average cost of about $30 per foot. To say that this has been an entirely useless expenditure of money would be hardly just; but it was certainly a premature and ill-advised investment of money, which was needed in other and more legitimate mining enterprises."

As will be seen from the above quotation, many of these tunnels were driven wildly for the purpose of prospecting. The practical use of tunnels, however, did not start until some time in 1888 and 1889, when power drills were introduced in mining practice. Shortly after the rather porous ore body at Aspen was found to be filled with so much water that the capacity of the pumps was taxed. The Cowenhaven tunnel was driven and reached the ore bodies 12,000 feet from the portal, near the bottom of the zone of secondary enrichment.

The second great undertaking of this kind was the Revenue Tunnel, at Sneffels, Ouray County, Colorado, which is 7,700 feet long and cut the Virginius vein 2,200 feet from the surface. A shaft was sunk from this level 750 feet, making it the deepest mining operation in the state.

The Newhouse Tunnel, at Idaho Springs, which is now 21,968 feet long, was started in January, 1904. This tunnel has been used for both drainage and transportation.

The Nelson Tunnel, at Creede, a crosscut 4,000 feet long, cut the great Amethyst vein 1,445 feet from the surface. This tunnel has been used for both drainage and haulage purposes.

On account of the surface conditions the problem at Cripple Creek was one of drainage only, and in 1906 the Cripple Creek Drainage & Tunnel Company was organized and work began May 11, 1907. Drifts from this tunnel reached Portland No. 2 shaft November, 1918, and the bottom of the Cresson shaft September, 1918. The distance from the portal of the tunnel to the Portland shaft is 26,370 feet. Work has now been discontinued on the main bore, but if continued to the Vindicator shaft, its first projection, the distance would be 27,190 feet from the portal.

The Camp Bird Crosscut Tunnel, from their mill, was started in December, 1914, and reached the Camp Bird vein May, 1918, a distance of 11,000 feet. This tunnel was driven for both drainage and transportation purposes.

The Yak Tunnel was driven for both drainage and transportation purposes and serves the mines on Iron Hill near Leadville. It is 1,200 feet below the surface in the Vega ground, which is at the heading.

It is to be noted that most of the long tunnels which have been driven for prospecting purposes have been failures. Tunnel-driving is very costly and should not be undertaken unless it can be proved that advantages exist either from decreased cost of transportation or pumping, which will repay the money invested.

ELECTRIC POWER

In the year 1906 construction of the system of The Central Colorado Power Company, predecessor to The Colorado Power Company, was commenced. This system was projected for the purpose of serving with electrical power the mining districts adjacent to its transmission lines, which extend through the mining districts comprising the so-called "sulphide belt." The power is distributed from two hydro-electric generating stations, one located at Shoshone, about ten miles above Glenwood, utilizing the water of the Grand River, and the other approximately four miles west of Boulder, utilizing water impounded in a large reservoir near Nederland. These two generating stations generate a maximum of approximately 40,000 horsepower.

Electrical energy is transmitted at a potential of 100,000 volts, stepped down and distributed from substations located at Leadville, Dillon, Idaho Springs and Denver. The secondary distribution lines serving the mining districts have a total length of approximately 325 miles.

The distribution of energy from this system was commenced in June, 1909, at which time the company was serving a load in the mining districts, representing some 2,000 horsepower. The intervening years have seen this load grow until at the end of the year 1918 there was utilized something over 20,000 horsepower for mining and metallurgical purposes. The mining manager was quick to realize the advantages of custom electric power service, evidenced by the fact that at the present time 95 per cent of all operating mines in districts served by the company are using electric power for pumps, mills, hoists, compressors, hauling systems, electrometallurgical furnaces and numerous other purposes which tend to simplify mining operations. The Colorado Power Company's lines have been extended into remote and isolated districts where operations under conditions existing a few years ago would have been impossible. There are sections served where the transportation of fuel, especially during the winter months, is a practical impossibility, and the ability to obtain electric power has enabled the operators to continue operations throughout the winter months, while heretofore the inability to obtain fuel would have caused the closing down of the properties.

During its existence The Colorado Power Company has signed agreements for service to over one thousand mining properties, a great many of which were prospects. Out of this number some large producing mines have been developed. Definite data is not available, but many of these properties have been added to the list of producers through being able to obtain electric service.

The mining districts served from the network of distribution lines by The Colorado Power Company are Boulder, Central City and Black Hawk, Georgetown and Silver Plume, Idaho Springs, Alma, Breckenridge, Dillon, Kokomo, Climax, Leadville, Red Cliff, Bonanza, Monarch and Garfield, and also extended areas in the Counties of Boulder, Gilpin, Clear Creek, Summit, Lake, Park, Chaffee, Eagle and Saguache. Energy is also delivered to Utah Junction, just outside of Denver, for use in electric furnaces manufacturing ferro-alloys and other metallurgical processes.

Principal mention is made of the extent of The Colorado Power Company's system, as this company supplies such a large mining area. A similar development is that of The Western Colorado Power Company serving the San Juan mining district, while The Western Light and Power Company and The Trinidad Electric, Transmission, Railway and Gas Company serve, respectively, the northern and southern mining districts of the state.

The Arkansas Valley Railway, Light and Power Company has an extensive distribution system serving the Cripple Creek, Victor and Canon City districts. It will be seen from this that the mining districts of Colorado have available a great amount of custom electric power, in fact enough to fulfill demands for some time to come.

These companies have made extensive study of the application of electric power to mining and metallurgical uses and have established rates which show a saving as compared with the cost of operating individual power plants. Properties requiring energy for electric furnaces and electro-metallurgical processes having high-load factors can obtain rates which place these industries in a favorable competitive position with those situated closer to the markets where the product is consumed. The advantages accruing as a result of the development of such a power system are apparent, and the availability of custom electric power has contributed in a large degree to the success and stability of the mining industry of Colorado; in fact, some of the largest mining districts owe their continued life to this one item. Formerly, with the high cost of operation and production, it was only possible to operate those properties producing high-grade ore, but now, with the advent of cheaper power and the improvements in metallurgical processes, those properties with lower grade ore are being profitably operated.

SMELTERS

The first smelters were located as near the mines as possible, but later, with the building of railroads, it was found advantageous to have them nearer important centers, like Pueblo and Denver, where operating costs were lower and a large ore supply was available. However, Leadville, Salida and Durango, on account of local conditions and the proximity of large mining operations, have continued to be important smelting centers. Other attempts to operate smelters near the mines have failed, and the constant improvement in concentrating mills has made it more economical to make concentrates, upon which both freight and treatment charges are lower than upon the raw ore. At the present time there are five smelters operating in Colorado with a daily furnace capacity of 4,000 tons, of which not more than 60% is utilized, although a large tonnage of ore is imported from other states. A further decline in this industry is to be looked for, unless there is a wide revival of mining throughout the state.

THE UNIT SYSTEM OF BUYING ORES

The purchase of ores is made on the basis of the unit. In the case of tungsten the basis is on the unit of tungstic acid, for molybdenum on the basis of the unit of molybdenum sulphide, for manganese on the basis of the unit of either metallic manganese or manganese dioxide. A unit may be defined as one per cent. of one ton or 20 pounds for a short ton, and 22.4 pounds for a long ton. If, now, manganese is quoted at 50c per unit metallic manganese, and we have a ton of 35 per cent. ore, the value of the ore will be represented by $35 \times .50 = \$17.00$. In other words, multiply the percentage of the valuable substance contained in the ore by the number of tons of ore—this gives the number of units in the ore. Multiply the number of units by the price per unit to find the value of the ore.

A more complicated case would be as follows: 7,200 pounds of tungsten ore has the following composition: tungstic acid 51%, phosphorus 0.05%, sulphur 1.0%, iron, manganese, insoluble constituents and water make up the balance. Tungsten ores containing 50 to 55 per cent. WO_3 are quoted at $15.00 a unit, and there is a penalty of $1.00 per unit for each 0.01% phosphorus contained over .04% and fifty cents per unit on every tenth per cent. sulphur over six-tenths per cent. contained. What is the value of the ore?

As phosphorus is limited to .04%, and there is .05% present, we have .05%—.04%=.01% excess phosphorus. As the penalty is $1.00 for each .01% over .04% we now have a penalty of $1.00 per unit, or the unit of tungstic acid is worth only $14.00. Similarly for sulphur we find the limit is .60% and we have 1.00%. 1.00%—.60%=.40% excess sulphur, but schedule calls for a penalty of fifty cents per unit of tungstic acid for each .10% over .60% sulphur, so .40% excess sulphur at fifty cents per unit tungstic acid makes a total sulphur penalty of $2.00 per unit, so that after subtracting this from our value per unit we have now a value of $12.00 per unit for our tungstic acid. Now 7,200 pounds of ore equals 3.6 tons. Applying the rule given above and multiplying by the percentage of tungstic acid in the ore we get

$$3.6 \times 51 = 183.6,$$

which is the number of units of tungstic acid for which the miner will be paid. The value of this is then found by multiplying by the corrected price (the price with the penalties deducted) per unit:

$$183.6 \times 12.00 = \$2203.20 \text{ value of ore.}$$

MINING IN COLORADO

The early history of Colorado is a narrative of the various expeditions into the Rocky Mountains in search of gold, and the first immigration to this region followed the actual discovery of placer gold in Cherry and Clear Creeks, near the present site of Denver, and subsequent discoveries near Idaho Springs: at Gold Hill, Boulder County, and at Blackhawk, Gilpin County, followed by vein-mining in all of the above named mountain districts. The outlying districts were explored, and as transportation, reduction and marketing facilities were improved, the regions where mineral had been found underwent vigorous development.

The total metal output of the districts now represented by our various counties is published for the first time in this report.

ARAPAHOE COUNTY

In 1858 and 1859 a small quantity of gold was recovered from the sands of the Platte River in this and Denver County. The earliest records of Colorado mining deal with the discovery of gold at the junction of Cherry Creek and the Platte by the Russell party. The amount of gold produced is said to have been about a hundred ounces. Various attempts to work the same gravels since that time have proved unprofitable.

ARCHULETA COUNTY

There has been at various times a considerable amount of prospecting in Archuleta County, and while it is said that large bodies of low-grade ore have been discovered and small quantities of high-grade, no mining operations of any importance have been undertaken.

BOULDER COUNTY

At almost the same time that discoveries of gold were made by the Jackson, Russell and Gregory parties in Clear Creek and Gilpin Counties, Captain Thomas Aikins and Charles Clouser opened a placer deposit near Gold Hill, Boulder County. Quartz deposits carrying free gold were discovered later, and the first quartz stamp mill in this region was constructed in the Gold Hill District, in the fall of 1859. Other mills went up and the placers were worked vigorously for a few seasons, and then little was done in the mines of Boulder County until 1869-72, when the rich veins of telluride ores were found. About the same time Sam Conger and associates discovered the silver mines near Caribou, and a general revival of mining occurred. The output of gold and silver reached almost $800,000 in 1879, and only in the year 1892, when the production was $1,131,767, has this been exceeded.

It was not until 1900 that the so-called "black sulphide," found in the neighborhood of the town of Nederland, was finally identified as ferberite (a tungstate of iron). About the same time tungsten metal was proving to be valuable as an alloy in self-hardening tool steel, so that a good market was soon open for the Boulder County ores. Up to the end of 1915 Boulder County had given the world tungsten valued at $5,000,000, $1,336,600 of which was produced in 1915.

For several years prior to 1915 the Jamestown District of Boulder County produced about 3,000 tons of fluorspar per year for the steel works at Pueblo. In 1917 the war demand for this material stimulated the production and about 15,000 tons were shipped to Eastern steel manufacturers in the next two years.

BACA COUNTY

A small production of a copper-silver ore has been made from the head of Carizzo Creek. The ore bodies are contained in the

Dakota sandstone and are small. The igneous intrusion at the Two Buttes was also the scene of some mining activity in the early eighties, but no production was made.

CHAFFEE COUNTY

The first actual mining done in Chaffee County was in 1859 and 1860, on the gravel deposits of Kelly's Bar, near Granite. Other placer operations in the neighborhood of Buena Vista and the mouth of Cottonwood Gulch were active at this time, and all continued until 1862, when the general exodus to the new fields began. Lode mining at the camp of Monarch got its first real start in 1881. In 1885 the production of gold, silver and lead exceeded $1,000,000, reaching $1,500,000 in 1887; and while the output has exceeded a value of $1,000,000 but one year since 1888, it has ranged from $200,000 to $800,000 annually.

The principal lode mines of this county are located at Monarch, Romley and Sedalia, a few miles above Salida. The iron deposits located at Calumet, which once produced a fairly heavy tonnage of iron ore, have been practically exhausted.

CLEAR CREEK COUNTY

The discovery of placer gold made by George A. Jackson, upon what is now the present townsite of Idaho Springs, January 5, 1859, was the real beginning of the mineral industry of Colorado. Many other diggings were opened the following summer and small fortunes were made. Soon, however, rich lodes were discovered, which made vein mining the more lucrative. Lead, copper and zinc were found in many of the mines in Clear Creek County, but it was not until 1885 that zinc was produced in commercial quantities. Careful estimates indicate that this county produced, prior to 1890, about $40,000,000 in gold, silver, lead and copper, but since 1890 the value of the annual production has slowly declined. The amount of zinc mined, however, has shown a substantial increase since 1903.

COSTILLA COUNTY

Gold was found in Costilla County as early as 1874, and placer operations were started in the Greyback District in the early eighties. Lode mining began soon after and has continued inter-mittently down to the present day. In 1882 The Colorado Fuel & Iron Company opened the Star of the West iron mine, but it was closed two years later, when the hematite of the outcrop gave way to pyrite as the mine was deepened.

CONEJOS COUNTY

At Platora, in 1871, gold ores were found, and since that time considerable development work has been done. The maximum production was in 1898 and was valued at $36,000.00

The large bodies of low-grade ores, which have been found in several of the well-developed mines, could be mined at a profit, if milling facilities were available and the cost of transportation to the railroad was lower.

CUSTER COUNTY

Attention was first attracted to the mineral possibilities of this county by the discovery of float quartz near Rosita Springs by Richard Irwin, in 1870. Prospectors were attracted to the district in great numbers and in the spring of 1874 Leonard Fredericks opened the Humboldt, and O'Bannion & Company located the Pocahontas. In 1877 the Bassick Mine was discovered, but was first called the Maine, and was located by John W. True, who was prospecting for a group of Pueblo people. It was abandoned in 1878 and relocated by E. C. Bassick, who made a fortune out of it.

The rich silver ores of Silver Cliff were first discovered by R. S. Edwards, in 1877. In 1880, soon after the Leadville excitement, which followed the discovery of the rich carbonates, this region was overrun with prospectors, and a veritable stampede followed. Many properties were opened and the shipping of ore commenced, but the oxidized silver ores failed to continue with depth and production started to fall off in 1881. Litigation finally closed the regular producing properties in 1884. The Bassick, however, again resumed operations in 1890, but closed the following year. Once more it was reopened in 1900 and made a relatively large production, but was closed in the spring of 1904. It is now being opened for the fourth time.

The ores near Silver Cliff, the scene of the early excitement, are mainly chlorides and sulphides of silver associated with lead and other carbonates. This old camp continues to ship to the smelters, and is showing signs of revival.

DOLORES COUNTY

The history of this county centers about Rico and practically begins in 1879, when Col. J. C. Hagerty found that some of the lead carbonates from Nigger Baby Hill were rich in silver. The Grand View Smelter was constructed in 1880, and in the fall of the same year started operations. On account of the difficulties of transportation, treatment charges were extremely high and progress was slow, until the advent of the Rio Grande Southern Railroad, in the year 1883. With transportation facilities, development was rapid until 1893, when the price of silver and lead declined rapidly. No district in the state was more seriously affected by this decline than Rico.

With the entry of the Rico Wellington Mines Company, controlled by Utah men, into this district, Rico started on a productive era, which has continued to the present time.

Some silicious gold and silver ore has been produced at Dunton, 16 miles from Rico, and a mill constructed for its treatment. Development work was continued through 1917, but no milling was done. All work was stopped in 1918.

DOUGLAS COUNTY

Along Cherry Creek and in the Bijou Basin small placer operations have been carried on since 1870. The gold is derived from the Dawson Arkose, but is not present in sufficient quantities to justify large operations. At West Creek a small chalcopyrite-sphalerite vein has been opened up, but it is still in the prospect stage. Tin ore (cassiterite) is found in association with samarskite (a uranium and tantalum mineral), topaz, phenacite, amazon stone and quartz at Devil's Head.

EAGLE COUNTY

The great rush to Leadville in 1879 led exploration into what is now Eagle County. Ores similar to those occurring at Leadville were found along the Eagle River and locations were made in the same year. Only high-grade ore would bear the transportation and treatment charges, and in 1880 a local smelting plant was erected which subsequently produced a large amount of lead bullion, but was driven out of business in 1882 by the advent of the Rio Grande Railway, which opened the Leadville market for these ores. When the lead carbonates were worked out, mining in this county declined.

Better zinc markets and improved milling processes caused a reopening of these mines for the heavy sulphide ores that they were known to contain, and a large output of zinc has been made since 1909. The tremendous tonnage of these ores now developed insures many years' operation. These properties contain much oxidized iron ore carrying from 12% to 18% manganese.

The quartzite contact in this same neighborhood has made a good yield of silicious gold ores, and promises to become productive again.

Iron pyrite, suitable for the manufacture of sulphuric acid, is found in great quantity in these mines. It is said to average close to 50% sulphur as broken.

Zinc production reached 28,000,000 pounds in 1916, but has declined somewhat since.

EL PASO COUNTY

"The Pike's Peak" rush followed the discovery of gold on Cherry Creek and many of the settlers came to the Pike's Peak District. Placer mines were opened on Fountain Creek, but did not yield profitable returns. In 1892 the gold mines of Cripple Creek were opened in El Paso County (but this district was soon incorporated in the newly created Teller County). In 1913 and 1914 small shipments of copper ore were made from the sandstones of the Blair Athol District, north of Colorado City.

FREMONT COUNTY

In 1870 a vein of silver ore containing some nickel was discovered at Cotopaxi, and mines on this vein have operated intermittently ever since. In 1895 the stimulus of the Cripple Creek boom was felt, and many prospects on gold-copper veins were opened up about White Horn and Cameron. The production has been unimportant, however, and reached its maximum in 1915, when it was valued at $54,303.00.

GARFIELD COUNTY

Excitement over the discovery of some ore in this county caused a rush of prospectors in 1870, but nothing materialized. 1909 witnessed another excitement about Goldstone, on West Elk Creek. The Grey Eagle was the only shipping mine and the production has been small.

GILPIN COUNTY

The discovery of gold by John H. Gregory, on May 6, 1859, in the Gregory lode at Black Hawk, was immediately followed by a wild mining excitement. During the summer of 1860, sixty stamp mills and thirty arastras were in operation between Nevadaville and Black Hawk, in what is now Gilpin County. In 1861 most of these plants were closed on account of the ores changing to sulphides and the difficulties of amalgamation which made it impossible to recover the gold values. Mining declined from 1864 to 1867, during which time the industry was being subjected to a periodical visit of the process mania, to which Colorado has been so susceptible. The completion of the railways in 1870 caused an immediate revival of mining and the construction of the first successful concentrators soon brought smelters, the first plant of this character in the state being erected at Black Hawk.

Some of the mines of Gilpin County yield silver, and in a small district this metal predominates. The greater part of the output has been, however, smelting ores which carry good gold values.

GRAND COUNTY

Prospectors have been active in Grand County since 1859, when it is said that a prospector by the name of Sandy Campbell discovered gold in the Rabbit Ear Range. Several producing mines have been developed since, but the maximum production for any year was in 1900 and was valued at only $3,775.00.

GUNNISON COUNTY

A prospector by the name of Fred Lottes discovered gold in the Tin Cup and Washington Gulch Districts during 1861, but little was done until 1879 and 1880, when there was a rush to this district second to none in the history of the state. The excitement of 1880 and 1881 is well remembered by old-timers of Colorado, but the exodus during the following year was almost equal to the rush and mining immediately declined.

A great deal of prospecting has been going on for several years in the Tin Cup, White Pine and Vulcan Districts.

The Irwin District, near Crested Butte, produced some high-grade silver ore; the White Pine District made a good output of lead carbonates carrying silver values, and the Vulcan and Spencer Districts produced some copper ores which carried high values in gold and silver.

This county has one of the largest graphite veins in the country and has made a considerable output.

HINSDALE COUNTY

The first real interest in this county began in 1874, and the population rapidly increased until 1879. The inaccessibility of this district prevented its development until the branch of the Denver & Rio Grande was constructed in 1889 and a marked revival of mining occurred. The general depression of 1893 again retarded advancement and the industry has declined until there is practically no mineral production made at the present time.

HUERFANO COUNTY

Prospecting began in Huerfano County in 1875, when large veins of low-grade ore were uncovered on Sierra Blanca. In 1912 uranium and vanadium ores were found at La Veta, but no production has been recorded. Some gold has been washed along the streams that rise in the Sangre de Cristo Range. The yearly value of the production of gold and silver has never exceeded one thousand dollars.

JEFFERSON COUNTY

Placer mining was begun in Jefferson County as early as 1860, but the bars did not prove at all lucrative, so were quickly abandoned. In 1884 copper mines were opened near Evergreen and prospectors soon discovered a number of veins, but no large operations resulted from this activity. Attempts have been made to mine fluorspar.

LAKE COUNTY

As the story goes, a man by the name of Kelly, one of the party of Georgians from Russell Gulch, Gilpin County, found his way into the Arkansas Valley in the fall of 1859 and discovered gold in the vicinity of Granite, Lake County. The site of this find is now known as Kelly's Bar. This discovery led to the gold find in California Gulch the following year. Kelly mining district was organized in March, 1860, and other Georgians followed, headed by Abe Lee. Near the spot where Leadville now stands, Kelly and his party met W. P. Jones and associates, who were from Iowa. On April 26th of that year, they found rich placer gold in California Gulch. The find proved to be extraordinarily valuable and adventurers came to Lake County from far and wide the following year. This, then, became the most populous spot in the territory of Colorado, for over 10,000 people resided in California Gulch that summer. Meanwhile other placers were located. Some in the neighborhood of Granite were very productive. California Gulch saw its best days in 1861, and in a few years the camp was almost deserted.

The placer production was light after 1866 and finally dropped to less than $20,000 in 1875, at which time the Painter Boy and other gold lodes were beginning to be operated profitably with stamp mills; and in 1877 the gold production increased to $55,000. In 1873 Lucius F. Bradshaw discovered lead carbonate by the accumulation of heavy sand in his sluice boxes. In 1874 W. J. Stevens and Alvinas B. Wood made further discoveries of lead carbonates carrying silver, and in 1877 California Gulch was again seething with prospectors. The camp was organized into a city and named "Leadville" on January 14, 1878. It has been said that Leadville was the livest town in the world in 1879. It had 15,000 people that fall and again Lake County became the most populous in Colorado. The railroad was completed in 1880, and Lake's silver production to 1884 exceeded $54,000,000. By 1887 it became a great smelting center.

LAS ANIMAS COUNTY

The little production recorded from this county is derived, Harry A. Lee, former Commissioner of Mines, believes, from the mines of New Mexico. Some gold has been washed from the sands of the Purgatoire.

LARIMER AND JACKSON COUNTIES

The copper deposits of Larimer County have long attracted attention. In 1899 the new camp of Pearl was opened, and in the next two or three years prospectors from the Grand Encampment District examined most of the northern part of the county. The most promising lodes have been worked intermittently since that time and have produced some copper and zinc, with a little gold and silver. There has been some activity about Waldron in the last two years.

LA PLATA COUNTY

Although the search for placer gold began in this county as early as 1861, little production is credited to it. Owing to the great interest in the adjoining counties, there was but little prospecting in the La Plata mountains prior to 1878. Some very rich lode deposits of unusual occurrence were found, and a production valued at $4,500,000 has been made, but within the past two years shipments have almost ceased. The discovery of high-grade telluride ore in the Neglected Mine in 1893 created a sensation, but the mine was closed before the end of the year and has made very little output since. The next find to attract attention was the discovery of the May Day in 1902, and the gold output climbed rapidly to its maximum in 1907. In 1910-11 the Valley View, a neighbor of the May Day, produced some extremely high-grade gold ore.

MESA COUNTY

In 1885 copper ore was discovered in the Unaweep Canyon, and from time to time there has been some activity in this district, but no mine of any importance has yet been found.

MONTEZUMA COUNTY

George A. Jackson, the discoverer of the Idaho Springs placers, began as early as 1873 to systematically prospect the so-called Baker Contact on West Mancos Creek. The veins found at that time were of too low grade to pay, and the district was soon abandoned. In 1895 there was an attempt to develop the "Doyle Contact," but this also proved a failure. Placer mines have been operated spasmodically. In the last three years there has been some activity on a substance that has been given the name of "Haller-ite." This, it is claimed, added to iron or steel greatly improves the quality.

MONTROSE COUNTY

The first deposits to attract attention to Montrose County were the placers of the San Miguel and Dolores Rivers, and each year a small value in gold has been won from them. In 1895 an attempt was made to work the rich copper deposits about Cashin. In 1899 the carnotite ores were first discovered and a history of the development of this industry will be found under the article headed "Uranium."

MINERAL COUNTY

In 1890 N. C. Creede located The Holy Moses Claim at Creede and shortly afterwards sold it for $65,000.00 to a group headed by David Moffat. In 1891 Moffat built a spur railroad to this property from Wagon Wheel Gap, and later in the same year the Last Chance and The Amethyst lodes were discovered. A stampede started immediately and Creede became famous over night. Mining activity rapidly increased, and in 1893 the county is credited with a yield of $4,150,946.00, mainly silver and lead. Then followed the general depression after the panic of 1893 and 1894; after which production rose again to $2,729,166.00 in 1898. But the bonanza surface deposits were soon exhausted, and as the mines were deepened the disposal of ground water became a serious problem. Drainage tunnels were constructed, but the yield gradually declined until in 1917 Collins and Wheeler opened up a silver bonanza in the Quintette, on Monon Hill. A large tonnage of fluorspar was also mined at Wagon Wheel Gap in the same year.

OURAY COUNTY

A. W. Begole and Jack Eckles crossed over from Cunningham Gulch, San Juan County, in 1875, and reached the present site of the town of Ouray, where they made some locations. The same year a large number of prospectors went into the Red Mountain District, and in the fall Andy S. Richardson, William Quinn, William Clark and others found their way to the Sneffels District, where they located some of the claims which later became famous producers; among them, the Virginius, Terrible and Atlas. The following spring brought a great influx of people from Lake City and other points.

It was not until 1896 that the great value of the famous Camp Bird vein was determined. This property has since become known as one of the great gold mines of the world, having yielded a gross production of $29,000,000, with net profits in the neighborhood of $18,000,000.

The Red Mountain District became famous in the late '80s for its rich copper-silver and lead-silver ores. The contact mines, lying to the northeast of the town of Ouray, also became famous for their yield of high-grade gold and silver ores.

In recent years the production has fallen rapidly because of the closing of the Virginius and Camp Bird Mines; the latter for the purpose of completing the 11,000-foot crosscut tunnel for drainage purposes. A substantial increase in the production for the coming year is expected.

PARK COUNTY

The gold hunters of 1859 explored what is now Park County, and the gravel bars along the streams flowing into South Park yielded them much gold in the following two years. It was natural while hunting for placer gold that some should be attracted to the lodes, and during the summer of 1859 many promising prospects were staked out in the Alma District. Silver discoveries were made in 1871, but were not extensively worked until 1872 and 1873, when the Moose and Dolly Varden Mines started making a production.

The rapid fall in silver prices in 1893 discouraged prospecting and development, and when the ore bodies then exposed were worked out, operations ceased. Little was done until the advancing price for silver in 1917 caused a renewed interest in the old silver producers. Many mills have been constructed in this county, but most of the ores have been shipped to the smelters without milling. A number of smelters were erected in the neighborhood of Alma, and those operating in the early days of the camp are said to have had a brief period of prosperity. The new mills are apparently doing successful metallurgical work.

PITKIN COUNTY

The first discoveries in this county seem to have been in 1879. Not much production was made, however, until 1887, when the Denver & Rio Grande Railroad reached Aspen. Then the great bonanza of the Mollie Gibson was mined, and by 1892 the value of the yearly production had reached $7,920,486.00. The enriched ore shoots were soon exhausted, however, and as the mines were deepened much water was found. Milling was started in 1898, but the value of the production steadily declined until 1912, when there was a temporary revival due to the operations of the Smuggler Leasing Company.

PUEBLO COUNTY

The production credited to Pueblo County is derived, in all probability, from prospects about Beulah, in the Wet Mountains.

RIO GRANDE COUNTY

James L. Wightman and associates discovered gold near the town of Summitville, Rio Grande County, in June, 1870. The first shipments of ore began in 1873; mills were constructed in 1874 and 1875. In 1875 the output of these mills amounted to $281,000, but the recovery of values was so low that operations became unprofitable, and in 1887 the district was practically deserted.

ROUTT COUNTY

Captain Way discovered placer gold near Hahn's Peak in 1864, but little was done until 1874, when 150 men were employed by The Purdy Mining Company, but without profitable result. There is said to be much gold in this neighborhood, and other attempts have been made to recover it, without success. The Park Range from the Wyoming line to Rabbit Ear Peak has attracted attention at different times, but only desultory mining and prospecting has been conducted. The veins carry sulphides of copper, iron, lead and zinc, with low values in gold and silver.

The Royal Flush Mine on Hahn's Peak has made a small output of a fair grade of ore and is being developed at the present time. A small output of copper has been reported from the Yarmony District in the southeastern part of the county; a 30-ton leaching plant was installed at Copper Spur in 1916, and about 30 tons of electrolytic copper produced, but financial troubles and litigation closed the plant early in 1917.

SAGUACHE COUNTY

Discoveries at the head of Kerber Creek in 1879 resulted in a rush to Bonanza in the following year, and in the next two years 4,000 mining locations are said to have been registered at Saguache, the county seat. By 1894 the yearly production of the mines had reached $400,000, but the decrease in the price of silver soon closed most of them. In 1917 and 1918 a revival in the interest in the district caused an increase in the production, and several new mines were opened.

SAN JUAN COUNTY

A party of adventurers, including Chas. Baker, after whom Baker's Park was named, were among those who visited California Gulch in the summer of 1860, and later in search of placer gold, explored what is known as San Juan County. They knew nothing about lodes, and after suffering many hardships in this inhospitable region, returned with little to show for their wanderings. Before 1871, however, this region was traversed by many other parties of gold seekers, and in that year some notable finds were made in the San Juan Mountains, but, owing to the isolated situation of this district, the yield of precious metals was comparatively small up to 1882, when the Durango-Silverton Railway was completed.

San Juan has produced some rich ores of gold, silver, lead and copper, and has made a fairly consistent output. A number of mills of moderate capacity have been operated for many years and the amount of concentrates shipped has been large. Some of these of later years have contained much zinc.

There is much undeveloped territory in the county, with innumerable veins of complex low-grade ore. Some shipments of hubnerite, the tungstate of manganese, have been made.

SAN MIGUEL COUNTY

In 1875 the first prospectors entered what is now San Miguel County. Early that year John Fallon made four locations, among them the Sheridan on the famous Smuggler Union vein. Mr. Wightman, an associate of Mr. Fallon, located the extensions of the Sheridan, but the following year his locations were jumped. The Smuggler, a fraction between the Union 'and the Sheridan, was located by John B. Ingram and it was on this claim that very high-grade ore was struck, and the first shipping began in 1876. The Mendota claim, reaching over the Divide to the Ouray side of the San Juan Mountains, was located in 1878 by John Donellan and William Evart. These completed the locations on the Smuggler Union vein on the Marshall Basin side, which in later years were consolidated into the Smuggler Union Company. A continuous output has been made from these properties since the beginning.

The old Belmont Mine, located at the head of Savage Basin, entered the list of large producers, and at last passed to The Tomboy Gold Mines Company, Ltd., but the values failed to continue with depth, and the operations were transferred in 1899 to the Argentine vein, and in 1910 to the Montana properties, located on the same vein.

The Liberty Bell, another famous producer, was opened for operation in 1896, and has made since a very large output.

Other important producers are as follows: The Alta, in Alta Basin; Black Bear, in Ingram Basin; The Caribou, Suffolk, Carbonero, New Dominion and The Butterfly-Terrible, in the vicinity of Ophir; the Cimarron, Japan-Flora, Valley View and Lewis, near Telluride, San Miguel County.

Large bodies of vanadium ore are located along the San Miguel River and are being treated in a mill at Vanadium. This mill is the only one of its kind in the state and has been worked continuously since 1909.

SUMMIT COUNTY

In the fall of 1859 Reuben J. Spalding, John Randall, William H. Iliff, James·Mitchell, N. B. Shaw, Balce Weaver and others (fourteen in all) left the Pike's Peak region and went through South Park over the Snowy Range into Blue River Valley. Not far from the site of Breckenridge they discovered placer gold, which they determined by a test to run from 13 to 27 cents per pan. Here they laid out claims, established a camp and started washing gold which proved profitable. Several other camps were started in the following year and the population grew until it numbered about 8,000 in the early 60's, and the gold production previous to 1867 is estimated at more than $5,000,000.

The first notable discoveries of silver in this county were made in 1868-9, other discoveries of sulphide ores bearing gold and silver had been made in the neighborhood of Robinson and Kokomo, and a heavy production resulted in 1881-2. Summit County contains the foremost placer area in Colorado. Dredging operations were started in 1901 and have operated profitably since that time. More of these dredges are being built and the production had been increasing until the depressing effects of the war caused a falling off in production in 1917 and a still further reduction in 1918.

The Wellington, one of the largest zinc producers in Colorado, is in this county. This mine started with about 1,000,000 lbs. production in 1901 and increased this to almost 20,000,000 lbs. in 1917.

What is said to be the largest known molybdenum deposit in the world lies in the western part of this county, near the Lake County line. This property is controlled by the Climax Molybdenum Company, which is now making an output said to be equivalent to three tons of 90 per cent. molybdenite concentrates daily— as much as is being produced from all other sources in the world.

TELLER COUNTY

Robert Womach first discovered gold at Cripple Creek in 1890, but it was not until 1891 (when he sent a piece of float rock to an assayer who reported that it went $250.00 in gold to the ton) that his discoveries were made known. In April and May, the same year, a number of prospectors from Colorado Springs visited the district and located claims. On the 4th of July, 1891, Winfield Scott Stratton staked the Independence, which soon netted him a fortune. The growth of the district from this time on was remarkable, and in 1892 the population had increased to over 4,000. By 1900 the mines of Cripple Creek were producing at the rate of $18,000,000 annually. High-grade ore was found at grass roots, and in some of the largest producers continued in great richness to a depth of 2,000 feet or more.

The production of this district to the end of 1918 is valued at $302,492,478. Rich specimens from this camp are found in every representative mineral collection in the world and in almost every small private collection. It is distinctively a gold camp (the ores bear very little silver), and has never made a commercial production of any of the common metals. The great Roosevelt Drainage Tunnel, driven at a cost of approximately $1,000,000, is described in another part of this report.

GOLD

Placers

Placer gold has been found in the gravel bars along the mountain streams in almost every mining district in Colorado, but in only a few instances has it been profitably produced. The production of the pioneer period was from deposits of comparative richness and invariably gave way to lode mining.

The improved methods of more recent times have made it lucrative to work the low-grade, deep gravel beds, and some of the most profitable mining enterprises of the state are the dredge operations of Summit and Lake Counties. Other districts are being investigated and increased production is expected.

Lodes

In pioneer days only outcrop oxidized ore bodies, yielding free gold, were worked. Most of these changed to sulphides when water level was reached, and a large part of the production now comes from mines yielding gold associated with silver and the common metals. The exceptions are the mines of the Cripple Creek District and a few in other parts of the state that carry tellurides or free gold in a highly silicious gangue.

Gold production from all sources has decreased in the past four years, due chiefly to the closing of mines, caused by the constantly increasing costs.

Value

Value of the production of gold in Colorado in 1915, 1916, 1917 and 1918:

	1915	1916	1917	1918
Lode	$21,721,634	$18,440,897	$15,078,224	$12,314,600
Placer	693,310	712,924	*651,000	*630,000

*Dredges in Lake and Summit Counties.

SILVER

History

Although for many years Colorado led all other states as a silver producer, it cannot be said to have operating at the present time more than one distinctive silver mine of importance, but many mines were worked for their values in this metal alone, prior to the depression following the demonetization of silver in 1893. The principal source of silver prior to 1879 was the ores of Clear Creek and Gilpin Counties, which were reduced in the pioneer smelters located at Blackhawk. After the discovery of the rich silver-lead carbonates in Lake County (which were also treated in local smelters), the output of this metal increased rapidly and remained relatively high until the oxidized lead ores above water level were exhausted, while in more recent years the production has come chiefly from sulphide ores, which contain silver associated with a number of economic metals.

The so-called sulphide belt, extending from Boulder County to the San Juan District, contains many promising prospects of ores of this character, and the future success of the metal mining industry of Colorado depends, to a great extent, upon efficient methods for the treatment of these ores on the ground. The flotation process can be used to separate many of these mixed sulphide ores into their constituent minerals, and in some instances, cyanidation may prove effective in extracting the silver and gold values.

Production

Mine production of silver in Colorado was 7,304,353 fine ounces in 1917, a decrease of 352,191 ounces from the previous year. The estimate of production for 1918 is 7,071,768 fine ounces, indicating a decrease of over 225,000 ounces from 1917.

Markets

The silver market began to feel the effects of war demand in April and May, 1916, and except for a brief recession during the summer of 1916, the. quotations averaged around 75 cents per ounce until June, 1917. Under the stimulus of foreign purchases the quotations for the third quarter of 1917 averaged above 88 cents. During the last quarter of the year the trade restrictions fixed by the government of India upon both imports and exports, and the refusal of China (acting under pressure from Great Britain) to purchase, caused a decline in quotations to 86.268 cents. The average quotation for the year was 81.417 cents per fine ounce.

It was realized early in 1918 that the Pittman Bill would probably become a law and that it would have the effect of fixing the price at its present quotation of 101⅛ cents per fine ounce and the quotation gradually rose from 88.716 in February to the government price in September and made an average for the year of 96.772 cents per ounce.

LEAD

History

Carbonates of lead carrying gold and silver in varying quantities have been produced in the Leadville District, in Lake County; along Park Range at the head of South Platte River and its tributaries, in Park County; at Monarch and St. Elmo, in Chaffee County; near Silver Cliff, in Custer County; at Kokomo and Robinson, in Summit County; at Aspen, in Pitkin County; at Red Cliff, Eagle County; and at White Pine, Gunnison County.

The carbonates of lead usually occurred near the surface in large masses that were easily removed. The early history of lead-mining may be said to mark the second epoch of mining in Colorado. The sulphide era was to follow when deeper mining was undertaken in the late '80s and early '90s, and has continued to the present day. It is marked, however, by the discovery of the oxidized zinc ores in the Leadville District in 1910, which led in some degree to the withdrawal of attention from sulphide mining. The principal source of lead ores, however, continues to be the mixed sulphides which exist in almost every metal mining district in this state. The output constantly decreased after the maximum production was made for their zinc and lead content in 1905 and 1906, until it received a new stimulus incident to the production of the oxidized zinc ores in 1910, 1911, 1912, and the establishment of more favorable markets. Competition for these ores stimulated production in other districts, notably in Clear Creek, San Juan and San Miguel Counties.

Production

Our production of lead has, in most instances, been derived from the same ores that carried silver, and the greatest output was made during the period of maximum silver production. The decline of production due to the exhaustion of the lead carbonate ores, had a depressing effect on smelting, and if it were not for the pride of the owners in keeping up the industry in this state, and certain advantages due to local conditions, which, in part, offset the cost of importing high lead-bearing ores from other states, less than one-half of the present production of ore would be smelted at home.

Mine production of lead in Colorado was 67,990,012 in 1917, a decrease from the previous year of 2,924,075 pounds. The estimated production for 1918 is 64,282,841 pounds, indicating a decrease of 3,707,163 pounds from 1917.

Markets

The lead market in 1917 fluctuated wildly, starting at 7.626 cents per pound f. o. b. New York in January, advancing to an average of 11.123 cents in June, and then reacting to 6¼ to 6⅜ cents at the end of the year. Early in 1918 the market commenced to show signs of recovery from the depression in the latter part of the previous year, and it looked as though prices must go higher when producers received intimation from the War Industries Board that no advances above 7¼ cents New York would be permitted. Consumers took advantage of this situation and withdrew from the market, causing the price to decline. Scarcity of the metal finally developed and an agreement was reached with the War Industries Board, which had the effect of fixing the price, and lead was sold up to 8.05 cents, New York market, in July, and continued at this price till the end of November. When the market was again set free the price declined to 5¼ cents at the close of the year. The statistical position of this metal is considered relatively strong and a reaction is looked for in the market.

ZINC

History

The first important output of zinc in Colorado was derived from sulphide ores, and prior to 1908 Lake County made over ninety per cent. of the production. The low price for spelter following the panic of 1907, and the exhaustion of the higher grade zinc-iron-lead ores, caused the production of zinc to decrease rapidly, until the oxidized zinc ores were discovered in 1910 and the record production of 106,000,000 pounds was made in 1912.

Improved milling and marketing facilities led to the development of mines in Summit and Eagle Counties and a large output of zinc-iron sulphide ores is being made from these districts. Recently San Juan County has become one of the leading producers; many other districts in the state have large undeveloped ore bodies of similar character, but perhaps of lower grade. The mixed sulphide ores which contain silver and lead and sometimes gold and copper in addition to the zinc, promise to be a more important source of zinc in the future.

Production

Mine production of zinc in Colorado was 120,315,775 pounds in 1917, a decrease from the previous year of 13,989,688 pounds. The estimate of production for 1918 is 88,641,748 pounds, indicating a decrease of 31,674,748 pounds from 1917.

Markets

The heavy demand for spelter developed by the war continued throughout the period under review. The supply, however, was sufficient to meet all demands, and the price declined to 7.51 cents per pound f. o. b. St. Louis at the close of 1917. During the early part of 1918 general curtailment took place, while the price reacted to 9.09 cents in September and made an average for the year of 8.15 cents per pound. At the end of 1918 producers were having much difficulty in finding a market for their ore. These influences no doubt caused the heavy decrease in production for this year.

COPPER

History

No district in Colorado has been productive of a distinctive copper ore. Leadville, however, leads with a total yield of over 90,000,000 pounds of copper derived from sulphides associated with other commercial metals. San Juan County ranks second, Gilpin County third, Ouray County fourth and Clear Creek County fifth. Small amounts have come from nearly all districts, and were, as a rule, derived from the mixed sulphide ores as in Leadville, but some very high-grade ore has been produced in the Red Mountain District of Ouray County, and in different portions of San Juan County. Montrose County made a number of shipments of ores rich in copper and La Plata County, a few years ago, produced some heavy iron copper sulphides carrying from 4 to 15 per cent. The largest tonnage of ores (valuable chiefly for their copper content) produced in recent years has come from Rico, in Dolores County, the Hidden Treasure Mine in Chaffee County, and the Vulcan and White Pine Districts in Gunnison County.

Production

Mine production of copper in Colorado was 8,122,004 pounds in 1917, a decrease of 502,077 pounds from the previous year. The estimate of production for 1918 is 6,423,919 pounds, indicating a decrease of 1,698,085 pounds from 1917.

Markets

At the beginning of 1917 the quotation for copper was 28¾ cents. The market declined 2 cents, then reacted to 30½ cents at the end of the month. The average prices for February and March were above 31 cents. The market fluctuated between 25 and 30 cents for the next six months and was then fixed by the government at 23½ cents for the last quarter of the year. The price for the year averaged 27.180 cents per pound, New York market.

During the first six months of 1918 the government price was 23½ cents, for the following five months 26 cents was allowed, and during December there was no market. The average for the year was about 24½ cents.

MOLYBDENUM

History and Properties

The name molybdenum comes from the Greek word "molybdaina" (lead), and was applied in ancient times to many different substances. Cronstedt, in 1760, first pointed out the difference between molybdenite and graphite, while eighteen years later Scheele first prepared molybdic acid.

Molybdenum is usually produced as a hard, dull, grey powder, nine times as heavy as water and nearly as hard as glass. It is also produced as a soft white metal which is readily filed and cut. When carbon is present, however, it becomes very hard. It may be drawn into wire, but its ductility depends much upon the amount of working it receives. Its melting point is about 2,550° centigrade, which is higher than all the metals, with the exceptions of osmium, tantalum and tungsten. It is readily soluble in nitric acid and like tungsten, oxidizes at a comparatively low temperature.

Molybdenum unites with oxygen to form a series of oxides, some of which have acid properties. The most common of these is molybdenum trioxide (MoO_3) which forms the molybdates such as sodium molybdate ($Na_3 MoO_4$). This ovide is unstable, however, and may be easily reduced to molybdenum dioxide (MoO_2), and finally to molybdenum sesquioxide (Mo_2O_3). Because of the complex salts formed, this irregular valance makes the chemistry of molybdenum rather complicated.

Minerals

There are three minerals of molybdenum which are important commercially. The first of these is molybdenite (molybdenum sulphide, MoS_2). It is a bluish black mineral looking very much like graphite, from which it differs in being more flexible and in giving a bluish streak. It is very soft. Flakes of molybdenite are usually hexagonal in outline.

Molybdite is a yellow powder that occurs with molybdenite. It is a hydrous ferric molybdate (Fe_2O_3, $3MoO_3$, $7\frac{1}{2}$ H_2O) and not a molybdic oxide, as has been supposed.

Wulfenite, a molybdate of lead, $PbMoO_4$, occurs in red, yellow or green tablets in some lead deposits. The crystal form is very characteristic and is usually that of a flat four-sided tablet with beveled edges. It is easily scratched with a knife and is brittle.

Ilsemanite, a complex oxide of molybdenum, is a blue to blue-black and soluble in water. It has been found in some sandstones in Utah.

Test for Molybdenum

Evaporate the final ground mineral to dryness with 5c.c. of nitric acid, then add $\frac{1}{2}$c.c of concentrated sulphuric acid and again evaporate. After cooling add, very cautiously, drop of water to the residue. If molybdenum is present, a brilliant blue appears.

Commercial Deposits

The distribution of molybdenum is very wide. It occurs in almost all types of rocks from basalts to granites. Commercial deposits occur either in very acid igneous rocks, such as pegmatites, or in association with oxidized lead ores.

Ores of the first type have always supplied most of the molybdenum of the world and will undoubtedly continue to do so. The molybdenum mineral (not molybdite) 'occurs in these deposits as either a part of the rock itself (i. e., is magmatic), or is deposited in veinlets in the rock, and seems to be due to the hot circulating solutions which were directly connected with the intrusion of the magma. The molybdenum sulphide is usually weathered in the upper part of deposits and the yellow molybdite is developed. The gangue minerals are quartz, orthoclase and pyrite. (In some cases chalcopyrite is present, but deposits containing this mineral are undesirable.) Of such a type are the deposits worked by the Climax Molybdenum Company, at Climax, and the Primos Chemical Company, at Empire. Here the developments are on bodies of alaskite, which contain about 1% molybdenum. The molybdenite at the surface has been changed at least partially to molybdite by oxidation. This is a disadvantage when the ore is to be concentrated by flotation, for it (the molybdite) is not saved, so that the value of the ore depends on the molybdenum present as the sulphide. This oxidation is then a process of secondary impoverishment, and the value of the ore will increase with depth (i. e., as oxidation decreases) unless a sulphide enrichment is present, and of this we have now no indication, and, furthermore, the chemical properties of molybdenite render it improbable of occurrence. The ore is in large bodies, is easily mined. and has been found amenable to treatment by flotation. Numerous other deposits exist in Colorado. Most of them, however, are of small size, but many of them of much higher grade. The foreign production from New South Wales, Queensland, Norway and Canada are from similar occurrences.

The second type of deposits are confined to lead deposits that have oxidized. It is quite probable that some galena carries molybdenum and that, with the oxidation of this sulphide, lead molybdate is developed. The chief deposits are in Arizona and Nevada, as at the Mammoth Mine in Pinal County, Arizona, where it has been worked as a molybdenum ore. Foreign occurrences are widespread, but the amounts of ore are very small.

Concentration and Smelting of Molybdenum Ores.

The concentration of molybdenite ores is carried out by means of the flotation process, to which it is peculiarly amenable. The average saving is said to be nearly 90%, and the ratio of concentration 100 to 1. It is this result that has made possible the exploitation of the low-grade deposits at Climax and Empire.

In the electric furnace this concentrate is now smelted directly into ferro-molybdenum, with a loss of not more than 20 per cent. of the contained molybdenum, part of which is recoverable from the slag. The metal may be obtained by the reduction of the oxide or sulphide with hydrogen, by the Goldschmidt process, or by reduction of the oxide with carbon in the electric furnace.

Recent Developments

Colorado has now become the most important molybdenum producer in the world. The production is derived from the properties of the Climax Molybdenum Company at Climax, Summit County, and of the Primos Chemical Company at Camp Urad, Clear Creek County. Construction was started on these properties in 1917, and the milling of ore by the flotation process in 1918. The Climax mill is equipped for milling 500 tons of ore daily, while that at Camp Urad handles 200 tons of ore a day. It is estimated that molybdenite can be produced for fifty cents a pound at these mines. Besides these large properties at least two hundred small prospects containing molybdenite have been opened through the state, many of which are worthy of investigation, while the mineral has been reported from almost every mining county. The present low price, however, does not encourage development.

The equivalent of 665 tons of concentrate, averaging 90% molybdenum sulphide, valued at $1,436,400, is the estimated Colorado production for the past year, and this figure may be compared with the world production of slightly under 250 tons for 1915.

The Market

The market for molybdenum ore is a restricted one and the price depends very much on the individual buyer. Most of the marketing is done privately. The following is the average monthly price in dollars per pound of 90% concentrate, as quoted by the Engineering and Mining Journal for 1917-1918.

	1917	1918		1917	1918
January	1.81	2.233	July	2.16	1.134
February	1.80	2.150	August	2.14	1.000 (b)
March	1.90	1.901	September	2.18	1.106
April	2.10	1.800 (b)	October	2.20	1.00 (c)
May	2.95	1.250 (b)	November	2.20	0.875 (d)
June	2.15	1.250 (b)	December	2.27	0.875 (d)

b Nominal.
c Estimated—no market.
d No regular market.

Uses of Molybdenum

Like many of the rarer elements, the amount of molybdenum used and the variety of uses to which it could be put depended upon the available guaranteed supply. No commercial organization is going to expend money in developing uses for a product unless a permanent supply of that material is in sight. Hence, with the development of the present large commercial deposits, the uses of molybdenum are increasing by leaps and bounds.

First, its use in steel. Fifteen years ago it was regarded as a harmful constituent in alloy steel, and considerable prejudice against it still exists among steel-makers. This is due to the delicate treatment which molybdenum steels require. There is, therefore, some difference of opinion as to the effect of molybdenum. Stewart[1] says, "Five to ten per cent of molybdenum introduced into steel raises the elastic limit, increases the tensile strength and gives greater toughness in addition to other properties," while Swinden[2] says the same effect is produced by one or two per cent., and that above this limit steels actually decrease in value. Molybdenum imparts properties, however, which much resemble those imparted by tungsten, but the amount of molybdenum required is about one-half as great. Tool steels containing chromium, molybdenum and carbon are tougher than similar "high speed" tungsten steels, but considerable trouble has been experienced with them, because there has been difficulty in judging the temperature of quenching. The alloy known as "iridium steel," which contains cobalt, tungsten, chromium, vanadium and molybdenum, is said to do 60% more work than the best tungsten steel. Some idea of the value of these new steels may be obtained from a statement that the speed of lathes has been increased from fifty feet a minute with the old carbon steel tools to five hundred feet a minute with the new high-speed tools. Molybdenum is also said to be used in the production of acid-proof steels and magnetic steels. The following list of uses will give some idea of the extensive application of these steels:

Gun linings.

Trench helmets.

Armor plate.

Rifle linings.

Armor-piercing projectiles (anti-tank bullets).

Acid-proof steels (in chemical factories).

Automobile frames and axles.

High-pressure boiler plate.

Steel wire.

High-speed tools (lathe bits, twist drills, etc.).

Permanent magnets.

An alloy with chromium and cobalt (stellite) is used in the manufacture of non-tarnishing cutlery and tools of great hardness. A copper molybdenum alloy is said to temper well. Metallic molybdenum is used to wind the heating units of electric furnaces, in electric contact points; as supports of electric light filaments; and in dentistry as a substitute for platinum. As ammonium

---

[1]Molybdenum—Mineral Foote Notes, p. 4, January-February, 1918.

[2]Carbon molybdenum steels, Carnegie Scholarship Memoirs, Iron and Steel Trust. (London) Vol. 3, 1911, pp. 66-124; and A Study of the Constitution of Carbon-molybdenum Steels, ibid., Vol. 5, 1913, pp. 100-168.

molybdate, it is used in the determination of phosphorus and lead, while the sodium salt has been used in making a blue pottery glaze and in the dyeing of silk and wool. The oxide Mo_2O_3 is used to preserve smokeless powder, which is to be used in the tropics.

URANIUM

History and Properties

Klaproth, a German chemist, isolated from pitchblende what he believed to be a new metal, which he called uranium. Sixty years later, the French chemist, Peligot, showed that this supposed element was really an oxide of uranium, while in 1892, Becquerel, another Frenchman, noticed that uranium salts affected the photographic plate, and this observation resulted in the discovery of a new property of matter—radio-activity.

Uranium is a hard, white, lustrous metal, that takes a high polish. It is 18.7 times as heavy as water and has an atomic weight of 238.5. It melts at more than 1,850° C., and can be boiled in the electric furnace. It is easily soluble in hydrochloric, nitric and sulphuric acids, and somewhat more slowly so in boiling water. It burns in oxygen at 170° C. with a bright flame, but will not tarnish in air. It is prepared by heating oxides with carbon in the electric furnace. Commercially, its compounds are prepared by roasting pitchblende with sodium carbonate and sodium nitrate and extracting the melt with water or by the treatment of carnotite with an acid, usually nitric acid or with an alkaline leach.

Chemically, uranium is a very active element, forming two oxides, UO_2 and UO_3, and a third oxide, U_3O_8, which is the combination of these two. From these are formed four series of compounds: The first are formed from UO_2, known as uranous compounds, and a typical example is that of uranium chloride, UCl_4; these are easily oxidizable and green to blue in color. The rest of the compounds are formed from the oxide UO_3—they are first, the uranates for example, K_2UO_4 potassium uranate; the second, the diuranates, for example, $K_2U_2O_7$, potassium diuranate; and the uranyl salts, example, $UO_2(NO_3)_2$ uranyl nitrate. The uranyl salts are the most common and are yellow with a distinct greenish fluorescence.

Minerals

The most important mineral yielding uranium is carnotite, which occurs as a canary yellow powder in the sandstones of the McElmo formation in Colorado and Utah, and also in Australia. It is a potassium uranium vanadate, $K_2O.2UO_3.V_2O_5.3H_2O$.

The next most important one is pitchblende, a complex mixture of the oxides of uranium with some lead and the alkali metals. It is dull black in color and has a specific gravity greater than that of galena. It occurs at several mines in Colorado in the Georgetown District.

Other minerals are autunite, a hydrous calcium uranium phosphate ($CaO.2UO_3.P_2O_5. 8H_2O$) which occurs in brilliant yellow flakes and torbernite in green flakes, a hydrous copper uranium phosphate $Cu(UO_2)_2P_2O_58H_2Q$. Samarskite, a tantalate and columbate of iron, uranium, and the rare earths, heavy and resinous brown, is mined occasionally. There are a number of other minerals that contain uranium, but they are very rare and are known from only one or two localities.

Deposits

We can class uranium deposits in two groups.

First, deposits formed in gash veins in association with sulphides. Second, deposits formed in sedimentary rock by circulating waters.

The first group is usually associated with large granitic intrusions and suggest, because of the presence of minerals that are deposited at low temperatures and pressures, that they are the last phase of primary mineralization. The uranium mineral is pitchblende, and it is associated with quartz, pyrite and chalcopyrite, and in the European occurrences with cobalt, nickel and silver minerals and also native bismuth.

The most important deposits of this type are at Joachimsthal, Bohemia; Johanngeorgenstadt, Annaberg and Schneeberg, Saxony, and at Bodwin and Dalcoath and South Tresavean, Cornwall, England. The pitchblende deposits of Gilpin County are similar. They occur in nearly vertical fissure veins in a Tertiary monzonite and in pre-Cambrian granitic rocks in association with quartz, pyrite, chalcopyrite, sphalerite and galena. The pitchblende is found in pockets, which are scattered along the veins, and for this reason the amount of uranium ore produced has been small. Some oxidation has taken place so that the pitchblende is often coated with an apricot-yellow powder, which is supposed to be a uranium sulphate, The mines that have produced the ore are the Alps, Belcher, Calhoun, German, Kirk, Mitchell and Wood Mines on Quartz Hill; the Pewabic Mine in Russell Gulch and the Jo Reynolds Mine, near Lawson.

Autunite and Torbernite occur in similar deposits at Guarda and Sabugal, Portugal, and also near Farina, South Australia.

By far the greater part of the world's uranium and radium are produced from the deposits of the second type, which occur in western Colorado and eastern Utah, where a carbonaceous sandstone, the McElmo formation, contains numerous pockets of the canary yellow carnotite. The average uranium content of this ore is between one and two per cent.

The sandstones are nearly horizontal, but are much faulted and particularly so in the vicinity of the richer carnotite deposits. Coaly masses are scattered through them, while quartz sand with a little zircon, the oxides of iron and the vanadium mica roscoelite make up the remainder of the rock. The carnotite pockets are usually above impervious clay strata, never far below the surface of the ground and are mined by shallow tunnels and open pits.

The origin of these ores is still a matter of speculation. It is known that all sandstones are very slightly radio-active, due, probably, to almost infinitely minute quantities of radium and uranium. The McElmo sandstones and, in some places, the La Plata are very much more radio-active than the ordinary sandstones, and from this fact Moore and Kithil (Bull. 70, U. S. Bureau of Mines) have drawn the following deduction:

"The results (tests of radio-activity of the sandstone) seem to indicate that the uranium was disseminated in the sandstone country rock and has been concentrated in ore bodies by the action of water, the "bugholes," in some cases, at least, acting as channels for the ore-bearing solutions."

Production and History

In 1899, Friedel and Cumenge, two French chemists, described a new mineral that they called carnotite. They had obtained it from Poulet and Voilleque, and it was said to come from the Paradox Valley in Colorado. As it contained uranium, it at once excited the interest of Madame Curie, who was then engaged in her earlier researches on radio-activity. In the following year Poulot and Voilleque began the production of the new mineral. By 1910 the deposits had been thoroughly explored and the Standard Chemical Company had become the leading factor in the field. At first most of the ore was shipped to foreign countries for refining, but by the opening of the European War in 1914, the American refiners were firmly intrenched and now, at the end of the most productive year in the history of the district, three American companies control the field. The Colorado Radium Company, the Standard Chemical Company and the Radium Luminous Materials Corporation. In 1914 the U. S. Bureau of Mines entered the field in connection with the National Radium Institute and operated some claims owned by the Crucible Steel Company until 1916. As a result of this work, better methods of refining were discovered. The production is given in the following table:

VANADIUM, URANIUM AND RADIUM PRODUCED IN COLORADO—1900-1918

*Radium in Grams	Uranium in Pounds	Vanadium in Pounds
1900............ 3	20,145 U. S.	?
1901............	Ore 375 Tons U. S.	....
1902............	Ore 3,810 Tons U. S.	....
1903............	Ore 432 Tons U. S.	....
1904............	Concentrates 44½ Tons U. S.	
1905............	Very Small Production	....
1906............	200 Tons Ore U. S.	600 tons Ore U. S.
1907............	?	21,600 lbs. U. S.
1908............	No Production	
1909............	No Production	20,000 lbs. U. S.
1910............ 6	42,400 U. S.	168,000 lbs. U. S.
1911............ 6	42,400 U. S.	168,000 lbs. U. S.
1912............ 6*	44,000 U. S.	60,000 lbs. U, S,
1913............ 8.5*	72,600 U. S.	864,000 lbs. U. S.
1914............ 22.3*	174,400 U. S.	904,000 lbs. U. S.
1915............ 6*	39,800 U. S.	1,254,000 lbs. U. S.
1916............ 7*	50,000 e	1,400,000 lbs. e
1917............ 7*	50,000 e	1,600,000 lbs. e
1918............ 14*	94,000 e	2,000,000 lbs. e

* Radium content of ores on basis of 1 gram to 3,000,000 grams of uranium.
 e—Estimated.
 U.S.—Figures from the United States Geological Survey.

Mining and Prospecting

The fact that the deposits of carnotite are small must be emphasized. If a claim produced a thousand tons of 2% ore, it is a good claim, for the ore is in pockets in a comparatively thin, flat-lying band of sandstone, and is usually less than fifty feet from the surface. Because of the shallow depth at which the deposits are found, the Standard Chemical Company has used diamond core and rotator hammer drills in prospecting. This work is said to have been very successful. Mining is carried on through shallow tunnels and open pits, while the ore is hand-sorted at the surface.

In prospecting for carnotite it should be remembered that the ore is not likely to occur far from the surface, and that it is not likely to be found away from the McElmo or its equivalent sandstones. The McElmo may be described as the first grey sandstone above the "Red Beds." It is also marked by the occurrence of brilliant green or chocolate shales.

Uranium ores have now been discovered over a wide area, which includes all of Montrose and Dolores Counties west of longitude 108° 30' and extends as far west as the La Salle Mountains and the San Rafael Swell. Northward, it is found along the White River in Rio Blanco County and along the Green River in Utah, and its presence is to be suspected wherever Triassic sandstones outcrop in the Great Basin.

Treatment of Uranium Ores

Three products are recovered from carnotite: uranium, vanadium and radium. The ore as it is taken from the mine is too lean to treat directly and is usually concentrated by grinding to 70 or 80 mesh, and then either settling out the silica in water and drawing off the fine powdery carnotite in suspension or simply blowing the carnotite out of the finely ground ore with a current of air. The concentrate is then treated either by fusion with soda or soda niter, or dissolved in some acid (c. f. Bull., 70 and 104, U. S. Bureau of Mines).

Uses of Uranium

Uranium is used in medicine in the form of the nitrate and salicylate as a remedy for gout and diabetes. In chemistry uranyl nitrate is used as an indicator and it is used also in the titration of phosphoric acid. On pottery its salts produce with lead salts fine glazes, that are yellow, orange, brown, dark green or sometimes black. Six-thousandths of a per cent. is said to produce a good color. The greenish yellow fluorescent glass which was used so much thirty years ago is made by adding twenty per cent. of uranium oxide to the glass. In dyeing, uranium salts have been used as a brown dye and also as a mordant. In photography it is used in producing sepia tints and also as an intensifier on weak negatives. German patents have been issued for its use in gas mantles, but this does not seem to have been particularly successful. The carbide of uranium has been used as a catalytic agent in the manufacture of ammonia, and is said to be extensively used in Germany for this purpose. The carbide is also used in the manufacture of sparking metal for automatic lighters. The iron uranium alloy ferro-uranium is added to steel to impart toughness and hardness. It is reported that a self-hardening steel superior to tungsten steel can be made in this way.

The Market for Uranium Ores

The price of uranium ores held fairly steady throughout the year at about $50.00 per unit of uranium oxide. The black oxide was quoted at $3.25 to $3.75 a pound, while sodium uranate was quoted at from $2.25 to $2.50. The buyers of crude ore are:

> The Foote Mineral Company,
>   Philadelphia, Pa.
> The Radium Luminous Materials Corporation,
>   Naturita, Colorado.
> The Colorado Radium Company,
>   Denver, Colorado.
> The Standard Chemical Company,
>   Naturita, Colorado.
> O. B. Wilmarth,
>   Montrose, Colorado.

Radium

Radium is always associated with uranium in a ratio which is very close to one part in three million. This comes about from the fact that uranium breaks down slowly but at a constant rate and forms radium, while radium in turn breaks down and forms lead and helium. The ratio given above represents the balance between the rate at which the uranium breaks up and the rate at which the radium breaks up. This atomic disintegration is known as radio-activity. This property makes radium commercially valuable, first, because it induces luminescence in certain salts, such as zinc sulphide; second, because it affects the tissues of the body, particularly in the case of cancer; and third, because it is of especial interest to science, as it offers a key to the problem of the ultimate structure of matter.

Technically speaking, radium is a white metal, but it is usually prepared as either radium bromide or chloride, which look rather like common salt. The main part of the production of radium salts during the past year has gone into the production of luminous paints, which have been used in the manufacture of luminous signs, tapes, instrument dials and similar articles. It is said that luminous tapes were laid to define pathways on the battle front and that every instrument in an aeroplane had a luminous dial. This accounts for the heavy production during the last two years.

Recent Developments

Regarding present day developments, Mr. John I. Mullen, of the Standard Chemical Company, writes: "During the year 1918, there has been mined in Paradox Radium Field, 8,450 tons of carnotite ore, averaging about 2% uranium oxide and 3½ to 4% vanadium oxide. Of this, the Standard Chemical Company has mined 6,800 tons, the Luminous Materials Company, 1,000 tons; the Radium Company of Colorado (Schlesinger interest), 350 tons; Geo. Picket and associates, 300 tons.

The Standard Chemical Company's headquarters and radium extracting mill is located at Pittsburg, Pa. The Luminous Materials Company's headquarters and extracting plant are located at Orange, New Jersey. The Radium Company of Colorado, headquarters and extracting plant, located at Denver. George Picket is associated with Dr. McCoy, whose headquarters and extracting plant are located at Chicago.

Standard Chemical Company employed an average of 200 men daily; the Luminous Materials Company gave work to 45 men daily; the Radium Company of Colorado employed 35 men; Picket and associates worked about 25 men.

The Standard Chemical Company has 375 mining claims in Colorado; the Luminous Materials Company has about 60 claims; Radium Company of Colorado has about 40 claims; Picket and associates have about 40. The Cummings Chemical Company, which has not operated here during the past three years, owns about 35 claims.

During the past three years, and especially during the past year, there has been practically no prospecting for carnotite properties. The writer knows of only four individuals that have spent any time prospecting, and I don't suppose they averaged over two months in the year looking for the rare metal. This is very probably due to the fact that the mining of carnotite ore is not very profitable for the poor man. The experience I have had, shows that in order to make anything in the mining of carnotite ore, one must go on through with it and sell the finished product, such as the radium, uranium and vanadium. The open markets pay very little for carnotite ore, and I might say that the small miner or prospector gets but little for the ore after he has it sacked, and even when the United States Government was buying ore they paid less for it than individuals did before the government went into the business. Because of this fact, mining of carnotite ore by the small prospector is rather discouraging.

The average haul from the carnotite field to the railroad is 60 miles. Cost of transporting carnotite ore to the railroad is 39c per ton mile. The roads are narrow, on steep grades, and have many sharp curves. Some of these curves have radii of 25 feet; in spite of this, there have been some eight or nine trucks operating during the past year, hauling ore from the radium fields to the railroads. If the roads are not improved it will never be any more economical to transport ore by trucks, than it will be by horsepower.

The production of carnotite ore is beginning to be a fixed mining proposition in Colorado. During the past three years, the Standard Chemical Company has drilled 52,000 lineal feet in testing this country for carnotite deposits. The largest pocket we found gave up 52,000 sacks (which is about 2,080 tons). This pocket was about one-quarter of a mile back from the rim and about 30 feet under the surface, the overburden being solid sandstone. I might further mention that this body of ore was found by a Sullivan D. P. 33 drill. With this machine the Standard Chemical Company has devised means to put down dry holes to a depth of 30 feet.

Of the above referred to 52,000 feet, about half of this footage was drilled by D. P. 33 drills, the holes varying in depth from 15 to 30 feet. 26,000 feet was drilled with diamond drills.

During the past year, I have noticed that the Luminous Materials Company and the Schlesinger interest are following the practice of the Standard Chemical Company.

For your information, I am pleased to advise you that the Standard Chemical Company is investigating the possibility of moving its radium extracting plant from Pittsburg to Denver. Before the European war, the question of chemicals was the big drawback, but certain changes have taken place in the chemical world during the past two years, and it looks to me as though we can get our extracting plant closer to our mines.

VANADIUM

History and Properties

Colorado produces most of the vanadium of the United States and in the world markets is the most important competitor with the vanadium derived from the Minasragra deposits of the Cerro de Pasco, of Peru. The element was first discovered by the Spanish chemist, Del Rio, in the lead ores of Zimapan, Mexico, but he believed it to be an impure form of chromium. It was later (1830) isolated by Sefstrom from the iron ores of Taberg, Sweden.

Vanadium is a greyish white metal and is usually produced as a powder which has a specific gravity five and a half times as great as water. It melts with difficulty (1,720° C.) and oxidizes slowly in air. It has been obtained by heating vanadium chloride in a stream of hydrogen, or by reduction of vanadic oxide with aluminum or vanadium carbides. Impure silver white crystals of vanadium containing 4% carbon have been made by heating vanadic oxide and carbon in the electric furnace.

While vanadium is fairly active chemically and belongs in the same group as chromium, arsenic and phosphorus, two series of unstable salts are formed with ordinary acids, those in which vanadium is trivalent, and those in which it is pentavalent. Examples of the first group are vanadium trioxide, V_2O_3, and vanadium trichloride, VCl_3, and an example of the second is vanadium pentoxide, V_2O_5. Vanadium forms two acids, however, vanadic acid, H_3VO_4, and metavanadic acid, HVO_3. These form salts which are sometimes quite complex and in some cases are of the alum type.

Most of the vanadium minerals are readily soluble in hydrochloric acid (muriatic) and give with it a rich red solution which changes to green upon the addition of water until the color finally disappears. Hydrogen peroxide causes the red color to reappear, and if zinc is added, the color will again change to green.

In the case of minerals of the type of roscoelite the mineral should be either fused in soda or the fine powder boiled for a long time in concentrated acid before making this test.

Minerals

While there are a large number of vanadium minerals, but four are of any great importance. The first of these is carnotite, which has already been described as a mineral of uranium.

Roscoelite, the second, is a complex silicate, which is directly related to the micas. Like them it has the perfect cleavage and pearly lustre, but it is green in color and usually the flakes are found in star-like groups, while the folia are quite commonly intricately folded. Its specific gravity is low and it is rather soft. It is a hydrous potassium, iron, magnesium, vanadium, aluminum silicate, $4H_2O.K_2O.(MgFe)O.(Al.V)_2O_3.12SiO_2$, and the pure mineral carries about 20% V_2O_3 (vanadium oxide).

Patronite is a sulphide of vanadium, VS_4, and is greenish black. It shows no sign of structure. At Minasragra it occurs mixed with bravoite, a nickeliferous pyrite, and quisqueite, a complex substance composed mostly of carbon and sulphur.

Vanadinite is a lead chloro-vanadate, $PbCl.Pb_4(VO_4)_3$. It occurs in red or yellow hexagonal prisms which usually have pits at the top. It is rather heavy and readily melts before the blow-pipe and yields a small lead button. It has been mined in Arizona and New Mexico as an ore of. lead. (See notes· on Commercial Deposits.)

Commercial Deposits

While vanadium is very wide-spread as a constituent of igneous rocks, the amounts are small, seldom exceeding five hundredths of a per cent. It is more common in the igneous rocks poor in silica. Hillebrand believes that it is associated with the ferromagnesian minerals and it is supposed to replace ferric iron and alumina. It also occurs in the ash of coal (Oklahoma) the ash of gilsonite (Peru) and the ash of grahamite (West Virginia).

Of the commercial deposits, the carnotite deposits of Western Colorado have already been described.

Of the Minasragra deposits at Cerro de Pasco, Ries[1] says:

"The area lies along the western edge of a broad anticline in Mesozoic sediments, that are intruded by many dikes of eruptive rocks. The chief vanadium deposit is in a fault fissure, the mineral patronite being associated with a black carbonaceous mineral known as quisqueite, but which carries no vanadium. The carbon compounds are along the vein walls, and the sulphide in the center of the vein. Oxidation has yielded green and brown vanadium oxides, and in this connection the vanadium has been dissolved and carried into the cracks of the neighboring crushed shale. The ore may. carry as much as 50% V_2O_5, but it usually runs from 20% to 30% V_2O_5, or about 40% when calcined. This deposit is a most important source of the world's supply, and much of the ore.is shipped to the United States."

Vanadite occurs in many of the oxidized lead deposits of the southwestern United States. These are usually mined for their lead content, but it is said that some ores at Cutler, New Mexico, were sold for their vanadium content. Ores of this type have one

[1]Mineral Foote Notes, May, 1918.

important disadvantage, and that is the occurrence of mimetite (which contains arsenic) with them. These deposits are due to the oxidation of lead sulphides under arid climatic conditions.

The roscoelite deposits of San Miguel County are the most important in the United States. They occur as a greenish sandstone in the brilliant red strata of the La Plata formation, where it outcrops along the San Miguel River. Here the roscoelite is found filling the spaces between the grains of the sandstone and occasionally it is replaced by mariposite, a complex chromium mica. These deposits, it has been supposed, are derived from carnotite by oxidation and solution, or from the basic dikes which cut the La Plata sandstone at this point. The veins which dip at a low angle toward the east are an irregular lens shape. The average content of the ore is about $1\frac{1}{2}\%$ U_2O_5, but selected specimens, have, upon analysis, shown as high as 8% vanadic oxide. It should be remembered in prospecting for these ores that they occur toward the base of the uppermost of the red bed formations.

Roscoelite is also found in association with the tellurides of gold (but not in commercial quantities) in Boulder County, and at Cripple Creek, in Teller County, Colorado, and at Kalgoorlie, Australia, and in the Mother Lode in California.

The Concentration and Smelting of Vanadium Ores

Vanadinite is easily concentrated on Wilfley tables as it is very heavy. Roscoelite offers a slightly more difficult problem, but it also may be concentrated by wet concentration methods.

The concentrates may be treated by several different methods —usually they are roasted with a flux such as common salt or soda and the melt lixiviated with hot water and the vanadium precipitated from the resultant solution by means of ferrous sulphate. The resulting ferrous vanadate is reduced by the Goldschmidt process or in the electric furnace to ferro-vanadium, and is so sold.

Uses of Vanadium

Vanadium is used mainly in the manufacture of steels. Ferro-vanadium is a very efficient deoxidizer and is used with ferro-manganese and ferro-silicon to deoxidize steel. The proportions used range up to 0.5%, of which about 0.2% is driven off as an oxide of vanadium. It adds very considerably to the tensile strength of the steel. High speed tool steels are made by the addition of larger amounts of ferro-vanadium, which increase the hardness of the cutting edge of the tool and also increase the temperature at which the tool edge will fail or "lead." This increases their cutting qualities and adds much to their life. They are, however, inferior to tungsten steel tools. The amount of vanadium in such steels runs as high as two per cent.

Chrome vanadium steels have a high elastic limit and are used extensively in automobile frames and axles. They compete with nickel steel here and recently automobile practice seems to lead toward their more general adoption as a substitute for the nickel chrome steels; this is said to be due to their greater freedom from surface blemishes. Chrome vanadium steels are used also in light armor plate.

. Vanadium is also used to increase the strength of cast iron. Aluminum vanadium alloys are light and up to 10% vanadium are malleable. Very hard alloys are made with 53% vanadium and 47% aluminum. Aniline black and some of the permanent inks contain vanadic acid. Vanadium salts are used to produce colored glazes in pottery, glass and porcelain, while vanadic oxide is added to rubber to increase its density and tensile strength, and to decrease its porosity. It also makes the rubber oil-proof.

In medicine it has been used with potassium chlorate under the name "Vanidin," and it is also said to have some effect on the growth of plants. In chemistry hydroxylamin and hydrazin are determined by the reduction of vanadic acid, while it has been employed as a catalyst (in place of platinum) in the contact process for sulphuric acid. In the manufacture of quinone from phenol it has served as a deoxidizer, while it has been found useful as a developer in photography. It is also used to make a substitute for gold bronze paint.

Recent Development

A maximum production was maintained all the year by the Primos Chemical Company, at Vanadium, and the product is said to have a ready market. The ore was derived in the main from the Fall Creek Mines near Saw Pit   Some vanadium was also refined from carnotite outside the state by The Standard Chemical Company and also by the other radium companies. Some prospecting for ores by diamond drills was done, but the results have not been disclosed.

History

The vanadium deposits at Vanadium, were first recognized in 1899. Desultory prospecting was carried on until 1904, when the Vanadium Alloys Company entered the field and built a mill at Newmire (now Vanadium). This mill was worked with indifferent success until The Primos Chemical Company took it over in 1909. Since then it has been worked continuously and with good results.

Statistics for the Colorado production will be found in the article on Uranium.

Markets

The price of ferro-vanadium fluctuated during the year and reliable quotations are hard to get, as this product was furnished by two companies. It ran apparently from $3.50 per pound of contained vanadium to $7.50, depending upon the grade desired. Quotations on other vanadium salts are also untrustworthy. The market was strong and all the available supply was rapidly absorbed.

TUNGSTEN

In speaking of the tungsten industry for the past year, Young[1] says:

"Under the stimulus of war necessities there was great activity in the industry in the United States, as well as in other countries, in 1918. The signing of the armistice found large accumulations of stocks of both tungsten concentrates and ferro-tungsten, on hand, the effect of which circumstance was to produce a chaotic condition among producers of tungsten ores and manufacturers of tungsten products. Prices fell, the market reflecting the uncertainties which attended inevitable readjustment to a peace basis. The maintenance of a lower price by Great Britain caused export ores, wherever permitted by licensing regulations, to seek the high-price markets of the United States. The result was a superabundance of tungsten ore imports in the United States, which caused the condition described by Charles Hardy in his report on the tungsten market, elsewhere in this issue. Before the Committee on Mines and Mining of the U. S. Senate, W. R. Ingalls, on May 10, 1918, made the following statement about tungsten:

'The supply of this metal is probably rather superfluous, production having been greatly stimulated in 1915-1916, when the price rose as high as $100 per unit, but before the end of 1916 excess of supply had reduced the market to $16. At present the price is about $20. The supply both from domestic and foreign sources increased from about 1,300 tons in 1914, to about 10,000 tons in 1916. The behavior of this market has not indicated that there is any need for encouraging production beyond what natural factors will effect.'

"The market conditions in the latter part of 1918 amply bore out the foregoing statement that natural factors were the only requirements for stimulating production. At present the market is dormant and the industry in a similar condition."

---

[1]Young, Geo. J.—"The Tungsten Industry in 1918," Mining and Scientific Press, Vol. 107, No. 2, p. 78, 1919.

If one refers to the production table given below, the effect of this condition can be seen.

The Boulder County output decreased materially and by the close of 1918 almost all of the producers had shut down. One producer states that it is not so much the low price of the metal, but the spottiness and irregularity of the veins that caused the abandonment of active operations. It is doubtful if production at the present price of tungsten can continue in competition with the cheaper ores from foreign countries.

Some production of tungsten ore (hubnerite) was made from Gunnison and San Juan Counties. Scheelite and wolframite have been found in Lake County, and wolframite in northern Fremont County. The Leadville tungsten minerals were so mixed with pyrite as to be practically useless.

In 1917 The Boulder Tungsten Production Company opened an electric furnace plant for the reduction of ferro-tungsten, but the plant was closed in January, 1918. The Tungsten Products Company operated their furnaces throughout the past two years and have also produced some tungsten powder, while the Black Metals Production Company have been making a tungstic acid of exceptional purity.

TUNGSTEN PRODUCTION		Tons	Per Unit	Value
1900	High grade ore, 63% WO_3	40	$1.30	$3,216.00
1901	High grade ore and concentrates averaging 65% WO_3	65	2.25	8,875.00
1902	Ores and concentrates	166	2.50	24,900.00
1903	Mainly concentrates	243	2.50	36,317.00
1904	Ores and concentrates averaging 55% WO_3	375	5.50	125,000.00
1905	Mainly concentrates	642	6.00	231,120.00
1906	Mainly concentrates	789	6.54	309,603.00
1907	Mainly concentrates	1146	8.83	573,643.00
1908	High grade ore 180	587		164,220.00
	Concentrates 407			
1909	Production on basis of 60% WO_3	993	$5.00-$9.00	391,160.00
1910	Production on basis of 60% WO_3	1221	6.50- 8.50	553,100.00
1911	Production on basis of 60% WO_3	730	4.50- 8.50	261,492.00
1912	Production on basis of 60% WO_3	775	5.60- 7.50	293,611.00
1913	Production on basis of 60% WO_3	953	$7.50	428,726.00
1914	Production on basis of 60% WO_3	630	$5.85-$9.00	252,000.00
1915	Production on basis of 60% WO_3	960	5.50-45.00	1,688,640.00
1916	Production on basis of 60% WO_3	2401	$34.15	4,919,649.00
1917	Production on basis of 60% WO_3	2707	20.00	3,248.400.00
1918	Production on basis of 60% WO_3	1910	16.00	1,833,600.00

ELECTRIC FURNACE PLANTS

In 1917 four electric furnace plants in Colorado were engaged in the manufacture of ferro-alloys. The two in Boulder produced ferro-tungsten, and the two at Utah Junction, ferro-chrome and ferro-manganese. By the end of the year one of the ferro-tungsten plants had closed down, but the other three continued in operation for the year 1918. Keeney[1] gives the following description of the Colorado plants:

[1]Robert M. Keeney,—"The Manufacture of Ferro-alloys in the Electric Furnace." Am. Inst. Eng., Bulletin No. 140, pp. 1322-1323, 1918.

"Electric Smelting Plants in Colorado"

Tungsten Products Co., Boulder, 800 kilowatts capacity, ferro-tungsten and ferro-molybdenum.

Boulder Tungsten Production Co., Boulder, 400 kilowatts capacity, ferro-tungsten.

Ferro Alloy Co., Utah Junction, 1,200 kilowatts capacity, ferro-chrome, ferro-tungsten.

Iron Mountain Alloy Co., Utah Junction, 3,000 kilowatts capacity, ferro-manganese.

Power for all of these plants is supplied by the Colorado Power Co.

Ferro Alloy Company's Plant

The plant of the Ferro Alloy Co., at Utah Junction, contains one 750-kw., three-phase furnace, to which power is supplied by three 250-kw. transformers, connected Δ/Y to give 129 volts on the furnace cables, the transformer ratio being 13,200 to 75 volts. The furnace operates with an actual voltage of 120 volts and produces about 3 tons of ferro-chrome per 24 hr. from 40 per cent. Cr_2O_3 ore. The furnace consists of a steel shell of circular cross-section, 8 ft. (2.43 m.) in diameter by 7 ft. (2.13 m.) deep, lined with magnesite, and having three vertical carbon electrodes 12 in. (30.48 cm.) in diameter.

A 450-kw. furnace, of the same size as the 750-kw. furnace, is also operated on chrome ore. This furnace is three-phase in appearance, but operates electrically as a single-phase furnace. There are three vertical carbon electrodes 8 in. (20.32 cm.) in diameter with a conducting carbon bottom. Power is supplied by three 150-kw., single-phase transformers, two of which have a ratio of 13,200 to 95 volts, and one a ratio of 13,200 to 100 volts. From each transformer, one lead goes to one vertical electrode, and one lead to the carbon bottom, so that the transformers are electrically independent. The furnace voltage varies from 90 to 95 volts.

At intervals, 150-kw., single-phase furnaces are operated on ferro-tungsten. These furnaces have one vertical graphite electrode 4 in. (10.16 cm.) in diameter, with a water-cooled steel bottom contact. The furnace shells are 4 ft. (1.22 m.) in diameter, and are mounted on trunnions, the slag being poured by tilting the furnace. All the electrodes are regulated by hand. The chrome furnaces are operated without roofs and the tungsten furnaces with roofs. The whole plant has a power factor of 90 per cent.

Iron Mountain Alloy Company's Plant

The plant of the Iron Mountain Alloy Company, at Utah Junction, contains one 1,200-kw., three-phase furnace, and one 1,800 kw., three-phase furnace, giving a total capacity of 12 long tons of ferromanganese per 24 hours.

The 1,200-kw. furnace is supplied with power by three 400-kw., single-phase transformers, having a ratio of 13,200 to 75 volts, and connected △/△. The voltage on the furnace is 72 volts. The 1,800-kw. furnace is connected to three 600-kw., single-phase transformers having a ratio of 13,200 to 75 volts, and connected △/△. The power factor of the plant is 85 per cent.

The furnace shells in both cases are 18 ft. (5.48 m.) long, 8 ft. (2.43 m.) wide, 7 ft. (2.13 m.) deep, and are lined with magnesite. The electrodes on the smaller furnace are of carbon, 17 in. (43.18 cm.) in diameter, and on the large furnace 24 in. (60.96 cm.) in diameter; in both furnaces, and they are threaded for continuous feeding. No roofs are used, and regulation of electrodes is by hand.''

The chrome used in the manufacture of ferro-chrome was imported from California, Oregon and Wyoming. The manganese came mainly from Colorado, but some was imported from Wyoming. The tungsten ore used was mainly ferberite from the Boulder District. The production of these plants is not known.

MANGANESE

In Lake and Eagle Counties are found oxides of manganese which carry oxide of iron and occur in bodies of considerable size. Lake County has made a continuous production of this material for use as a flux in the smelters since the beginning of the smelting industry in the Leadville District. More recently it has produced large quantities for shipment to smelters at Salida, Pueblo and Denver. In addition, a large tonnage was shipped to the Colorado Fuel & Iron Company at Pueblo for steel manufacture, while in 1917 and 1918 a considerable tonnage was produced for shipment to eastern steel works. The manganese content of the Leadville product ranges from 10% to 40%; the iron content from 10% to 30%, and silica from 5% to 20%. The shipments to out-of-state points carried a specification for 26%, minimum, manganese content. All of this ore carries more or less silver and under normal conditions the producer would ship either to the smelters for the silver value, or to the steel works for manganese value, whichever was to his advantage.

The production of Eagle County was used by the local lead smelter from 1880 to 1882, and subsequently was shipped to the steel works located at Pueblo. The Eagle County deposits yield a product which carries from 14% to 18% manganese; 30% to 35% iron, and 2% to 5% silica.

Some small shipments of manganese dioxide suitable for chemical purposes have been made from western San Miguel

County, but the difficulties of transportation to the railroad and the limit to the size of the deposits, made production unprofitable.

Gunnison County has produced some 45% manganese ore which was sold to The Iron Mountain, Alloy Company for metallurgical purposes, and was reduced in their electric furnace smelter located at Utah Junction, near Denver.

The output of manganese exported from Colorado in 1917 was estimated at 5,000 long tons (dry weight), and in 1918 at 30,000 tons, while there was shipped to the Colorado Fuel & Iron Company at Pueblo 115,000 tons in 1917 and 125,000 tons in 1918.

MINOR METALS

Bismuth

Despite the high price of this metal, as far as is known, none was produced in Colorado during the past biennial period. Deposits are known, however, at Leadville and in the La Plata Mountains. A small pocket of silver-rich bismuthinite was mined in Boulder County several years ago, while small amounts of bismuth are often reported from the mines of the San Juan area and from Park County.

Antimony and Arsenic

The increase in the value of these metals, due to war conditions, led to an increased production in the United States. It is probable that small amounts originating in Colorado were produced from the crude copper, that was refined in eastern electrolytic plants and that some was recovered from smelter flue dusts. The "grey copper" ore of San Juan and Gilpin counties contains notable amounts of both arsenic and antimony. Stibnite (antimony trisulphide) is reported from many districts, and a small vein has been mined in San Juan County for its silver values.

Selenium

This metal is recovered from the anode muds of certain electrolytic copper refineries and from smelter flue dusts. The blister copper from the Colorado region seems, however, to be low in selenium, but the vanadium ores of southwestern Colorado contain a considerable amount[1] which is not saved at present.

As its ability to conduct electricity is affected by light, it is used in devices that light automatically, in photometric apparatus, in sound detectors and to make ruby glass.

[1]Gale, H. S.—Carnotite and associated minerals in Western Routt County, Colo. U. S. Geol. Survey, Bulletin 340, p. 261, 1908.
    Hillebrand, W. F., Merwin, H. E., and Wright, F. E.—Hewettite, metahewettite and Pascoite, hydrous calcium vanadates. Am. Philos. Soc. Proc., Vol. 8, pp. 34-35. 1914.

Tellurium

A little tellurium was recovered in the past year from anode muds, some of which was probably from Colorado. If a good use could be found for tellurium the state undoubtedly could produce several tons annually. Native tellurium occurs in the Vulcan Mine in Gunnison County in notable amounts.

Hess[1] estimates that up to 1914, 431 tons of this metal have been mined in the ores of the Cripple Creek District alone.

Platinum

This metal has been reported many times as occurring in Colorado, but none of the reports have been substantiated. In 1915 the writer received a very tiny flake of one of the platinum metals which was said to come from a gold placer in the Castle Rock area, but efforts to follow the matter up proved unavailing. It has been rumored that the pyroxenites of the Caribou District contain a small amount of platinum.

Cadmium

Some cadmium was recovered by the smelters from zinc ores originating in Colorado. It has been reported from the zinc ores of Leadville.

Cobalt and Nickel

These are not produced in the state, but have been reported on Grape Creek, Fremont County, and at Silver Cliff, Custer County.

Titanium

None is mined in Colorado. However, large bodies of titanic iron ore have been described by Singewald.[2]

Zirconium

The increasing use of zirconium in steels has led to a lively interest in its ores. None are mined in Colorado, however, but there is a promising deposit of zircon at St. Peter's Dome, El Paso County.

---

[1]Hess, F. L., Tellurium, U. S. Geol. Survey, Mineral Resources of the United States, 1914, Vol. I, p. 976.

[2]Singewald, J. T., Jr. The Titaniferous Iron Ores of the United States; Their Composition and Economic Value. (Bulletin 64, U. S. Bureau of Mines), pp. 125-140, 1913.

Chrome

No chrome iron ore is known to occur in Colorado. Some was imported from Wyoming and California for use in electric furnaces for the manufacture of ferro-chrome. A small boss of basalt near Boulder contains some picotite (chrome spinel).

POTASH

The success of the potash plants in southwestern Nebraska turned the attention of many people to the alkali lakes of eastern Colorado. Many of these were investigated, but none seem to have been discovered that were worthy of commercial exploitation. An attempt is being made, however, to extract crude potash salts from the brines of some lakes in the San Luis Valley and in South Park. No production is reported.

The phonolites of the Cripple Creek District have excited some interest, not only because of their high potash content, but also because there are many thousand tons of finely ground material available in the mill tailings. The United States Bureau of Mines has been working on the problem of the treatment of these fines, and hope to develop soon a commercial process for the extraction of potash from them. A description of these rocks will be found in Professional Paper 54 of the United States Geological Survey.

Deposits of alunite at Rico and in the Rosita Hills are now being investigated. It is doubtful, however, if any of these projects can produce potash in competition with the potash from the Stassfurt deposits or those of Alsace-Lorraine.

PYRITE

Sulphuric acid is one of the chief ingredients used in the manufacture of high explosives, and is therefore a war-time necessity of primary importance. It is manufactured from sulphur dioxide made either by burning crude sulphur (brimstone) or by roasting pyrites. Before the war, pyrite, to supply the sulphuric acid plants of the seaboard was brought by ship from Rio Tinto, Spain, where it was mined very cheaply and upon which the freight amounted to very little, as it was shipped as ballast. When, however, the submarine campaign assumed serious proportions it became necessary to divert the shipping used in this trade to other service and the manufacturers of sulphuric acid had to look elsewhere for their supply. Strenuous efforts were made by the Bureau of Mines and the United States Geological Survey to develop pyrites deposits in the United States.

An examination of the Colorado deposits revealed many of great promise, among which may be mentioned those of Leadville, Red Cliff and several small deposits in Gilpin and Clear Creek counties. Other deposits of considerable size exist in the San Juan country. The problem, however, soon resolved itself into one of freight rates, and it was found that pyrite of the highest grade could not well compete with the eastern pyrites. The freight rate between the mines and Chicago was not less than $6.00 per ton, which would mean a sale price of at least $10.00 per ton f. o. b. Chicago. Eastern pyrite of similar grade could be supplied at a lower figure, as could also pyrite from Wisconsin and the Illinois coals. Leadville and Red Cliff pyrite were supplied, however, to the sulphuric acid plant of the Western Chemical Company, at Denver. The following is from a paper by Howard S. Lee[1] (died of influenza in the late autumn of 1918) :

Method of Mining

"In the past, the method of mining has been to drive exploratory drifts to locate the zones of enrichment, and the walls. As soon as this preliminary information was obtained, stoping was started, using the ordinary square-set method with selective mining. If it was found too expensive to hold the more or less barren pyrite in place, it was taken out and sold to the smelters for what little gold or silver it might contain, as well as for the iron. In two or three cases, notably at the Yak and Leadville Unit, it was found more profitable to mine the entire ore body, since the zones of enrichment were small but frequent, and the mass was quite heavy. In this way the unit costs were reduced, the tonnage increased, and the total profit returned was greater than if the richer parts had been gouged out and the larger part allowed to remain in place.

Value in the Manufacture of Sulphuric Acid

Because of the small amount of sulphuric acid manufactured and consumed in Colorado, and the high freight rates to the eastern part of the country, which precludes competition with Eastern and Spanish pyrites, no great amount of Leadville pyrite has been used in the manufacture of acid, although small shipments have been made for a number of years to the Western Chemical Manufacturing Co., of Denver, and to E. I. Du Pont de Nemours & Co., at Louviers, Colo. Occasional trial lots have been shipped East, which have roasted satisfactorily. Unfortunately, it is not the policy of the Western Chemical Manufacturing Co. to give out information concerning their practice, so that no comparison of methods and results with Leadville pyrite and with Spanish pyrite can be made through this company, but the fact remains that they have used and are still using a good deal of Leadville pyrite satisfactorily.

[1]Howard S. Lee, Pyrite Deposits of Leadville. Bulletin No. 140. American Institute of Mining Engineers, pp. 1223-1228.

"The following information was given by D. S. Robinson, Superintendent of the Du Pont plant at Louviers, Colo.:

" 'Our experience with Eastern pyrite has been somewhat limited, but, as a general statement, we would not hesitate to say that the Leadville pyrite roasts fully as well as Eastern pyrite, both as to being free running and in ease of obtaining almost a dead roast. Our ore is crushed to a maximum size of ¼ in. diameter. The small amount of sulphur left in the cinder is contained in those lumps which do not disintegrate to any extent during roasting. The burners are the five-hearth wedge type, with central shaft and stirring arms or rakes. The pyrite maintains its own combustion.

" 'We have burned pyrite from a number of mines in the Leadville District, and in one case only was the arsenic high enough to interfere with our process for the manufacture of sulphuric acid. Any arsenic in excess of 0.2 per cent. makes the ore unfit for the manufacture of sulphuric acid as carried on at our Louviers plant.'

"All pyrite shipments from Leadville in the past have been run-of-mine and no attempt has been made to deliver a sized product. The ore was crushed to suitable size for sampling and subsequent roasting in 'fines' burners. For this reason, no reliable information is available concerning the roasting qualities of lump pyrite. The writer made certain tests in an assay furnace with lumps as large as 6 in. in diameter, with which more or less violent decrepitation occurred, particularly in the larger lumps. This may have been due to excessive heat and to the fact that the heat was not applied gradually. It is also probable that an excessive amount of fines would be made during the handling of lump ore. At any rate, larger experiments under proper conditions are necessary before anything definite can be said regarding the value of Leadville pyrite as lump ore.

"During 1917, about 2,000 short tons of run-of-mine pyrite were shipped to the Western Chemical Mfg. Co. from the Tip Top mine, the average sulphur content of which was 48.31 per cent. Since the ore was sold for sulphur only, no further analysis of the control samples were made, but a number of mine samples were analyzed, showing less than 1.0 per cent. zinc, trace of lead, less than 0.1 per cent. arsenic, and 1.5 per cent. insoluble. During 1918, 5,000 short tons of pyrite were shipped to Denver and to smelters from the Denver City and Quadrilateral claims. The average contents of this ore was:

	Per Cent.		Per Cent.
Sulphur	48.0	Arsenic	Trace
Iron	45.0	Lead	Trace
Manganese	2.0	Lime	Trace
Insoluble	2.2	Moisture	1.24
Zinc	0.6		

"Two lots, approximately 100 short tons each, contained 50 per cent. sulphur, and one lot contained 51.2 per cent. sulphur.

".At the Greenback, Louisville and Yak, several carloads of pyrite have been shipped to various acid plants, containing from 42 to 49 per cent. sulphur. The ore in the southern part of the district contains more zinc and insoluble than that in the northern part, but not enough to interfere with its use in the manufacture of sulphuric acid, provided care is exercised in mining to eliminate portions containing zinc.

*VALUATION OF $5 AND UNDER

Per Ton
(2,000 Lbs.)

Leadville to Colorado common points.......................$1.50

(Colorado common points are Denver, Colorado
Springs and Pueblo.)

*VALUATION OF $20 AND UNDER

Colorado common points to Chicago	$4.45
To Mississippi River common points	3.65
To Memphis	4.45
To Nashville	6.10
To Pensacola	4.00

"It has been a general practice with local acid manufacturers to sell calcined residues to Colorado smelters. This could be done profitably because the smelters have always had an excess of silicious ores and iron has commanded a premium.

"Unfortunately, there is no place in the Middle West where residues can be sold, and if there is any metal value in the residue, it must be returned to Colorado smelters for treatment. This is now being done with certain residues from Kansas and Oklahoma zinc smelters. An average pyrite containing 40 per cent. iron in excess of silica, and 45 per cent. sulphur, will roast with a loss of 30 per cent. in weight and an increase in the iron to 56 per cent. The proportion of silver and gold in the residue will also increase in the same ratio. The iron credit will pay the treatment charges, so that the net cost of treatment is the return freight to Denver,

---

*See report of Smelter and Ore Sales Investigation Committee, State of Colorado, 1917.

Salida or Pueblo. Since all Leadville sulphides contain some gold and silver, the sale of residues is to be considered as entirely practical, where the original ore contains 4 oz. or more silver, or its equivalent in gold.

Summary

"There are large deposits of pyrite in Leadville, containing a minimum of 43 per cent. sulphur, suitable for the manufacture of sulphuric acid. In these deposits, there are enrichment zones containing gold, silver and zinc, but the system of mining in use will permit selective methods whereby the portions more valuable for other metals may be stoped separately and shipped to the smelters, while the pyrite is kept clean for acid use.

"The only material available now is run-of-mine ore suitable for crushing and roasting in 'fines' burners. The problem of lump ore, both as to suitability and preparation, will be worked out only when there is sufficient demand to warrant experiments.

"The returning of residues from a pyrite originally containing precious metals to the value of, say, $2 per ton, after calcination in the Middle West, for smelting in Colorado, is financially practicable."

Sulphur has been found in a number of counties in the state. The most prominent deposits are 25 miles southwest of Creede, on Trout Creek and Middle Fork. They are due to the action of hot springs on a tuff. Similar deposits occur in schists at Vulcan, in Gunnison County, where the sulphur is associated with native tellurium and rickardite, a copper telluride. Some sulphur has also been found in the entrance of the Black Canon, above Delta, while another deposit of some promise was found near Victor. As far as is known, the only sulphur shipped was a few cars from the deposits on Trout Creek.

FLUORSPAR

Location and Production

The flourspar deposit, located at Wagon Wheel Gap, Mineral County, has been worked continuously since 1915. Prior to 1917 the entire output was marketed within the state, going to the Colorado Fuel and Iron Company, at Pueblo. The production for 1918 is estimated at 15,000 short tons.

A considerable tonnage of fluorspar was shipped from the Boulder County deposits, located in and around Jamestown, but no steady output maintained. The grade of the crude ore was undesirable, however, and late in 1917 some milling was done with good results. Commencing about mid-summer, 1918, a much heavier tonnage was produced and shipments of marketable product for the year was estimated at 10,000 tons.

In the neighborhood of St. Peter's Dome, El Paso County, on the Cripple Creek Short Line Railroad, occur a number of fluorspar veins, some of which were opened up and the product shipped. On account of the intimate association of the fluorite with quartz, the crude material was apparently unmarketable, and it is not amenable to ordinary methods of concentration.

In some of the metal mines of Ouray County, fluorspar is found in commercial quantities, notably at the Barstow Mine, in the Red Mountain District. This mine produced in 1918 about 1,700 tons (averaging above 90% calcium fluoride) that is said to have been the most desirable spar shipped from Colorado. The Torpedo Eclipse and Ruby Trust Mines, at Sneffels, contain good grades of fluorspar, and discoveries have also been reported in other metal mines in this county.

A number of shipments of fluorspar were mined from a small vein near Evergreen, in Jefferson County, but the quantity of silica present made the product objectionable, and after a few shipments, operations ceased. Other finds have been reported, but no spar from there was marketed.

Small deposits of fluorite have been opened in Custer County, and a number of cars of very desirable grade were produced, but the bodies were irregular and the price paid did not warrant further development.

A few veins of fluorspar in Park County (one in the neighborhood of Bishop's Ranch, near Lake George) have been developed and some shipments made, but again the market price was not sufficient to justify production. There are other deposits in this county in the vicinity of Granite and Jefferson, on the South Park line of the Colorado and Southern Railroad.

Small deposits occur in Gunnison, San Juan and La Plata Counties, but there have been no commercial developments reported.

Market

The only property furnishing any appreciable quantity of commercial grade as a mine run product, was the mine located at Wagon Wheel Gap, in Mineral County. This deposit was vigorously developed and the production pushed up to 80 tons daily, when the demand was greatest, about mid-summer, 1918.

The district near Jamestown, Boulder County, had made shipments at various times for a number of years, but the grade of spar produced was low and contained excess deleterious substances. Some local operators finally undertook to dress the crude material by gravity concentration and results proved to be fairly satisfactory. A price was finally agreed upon, which was com-

mensurate with the costs and a considerable quantity of concentrated product was shipped in the latter half of 1918, and it is
said that some contracts were made for a portion of this production
extending through the year 1919, at prices which would leave the
producer a fair margin of profit.

It is said that about the same situation obtains with the Wagon
Wheel Gap production. One of the owners of this property stated
recently that they had a contract which would carry well into 1919
at a price somewhat above $20.00 f. o. b. cars at the mines. There-
fore, it would appear that the market as established in the last half
of 1918 for Colorado fluorspar of commercial grade, ranged between $20.00 and $25.00 per short ton f. o. b. cars Colorado. The
market in the future, and particularly after contracts begin to
expire, may be expected to decline, but to what extent, of course,
will depend upon supply and demand, competition with foreign
product, and all other market factors which tended to rule the
price under pre-war conditions.

The market price on the imported product per short ton
f. o. b. Atlantic seaboard, according to Ernest F. Burchard, of the
United States Geological Survey, is about as follows:

1909	$3.78	1914	$3.82
1910	3.18	1915	3.19
1911	2.46	1916	4.38
1912	2.74	1917	8.50
1913	3.15		

To this price on foreign product should be added $1.50 per
ton, import duty, which would make the price for 1916, $5.88, the
last year in which any considerable quantity was imported.

The quality of the Colorado fluorspar, produced either as
crude mine-run at Wagon Wheel Gap, or the mechanically dressed
product from the Jamestown District, is much superior to the
imported product, hence it would appear reasonable to expect a
differential in favor of the Colorado material. The high cost of
ocean shipping and constantly increasing use of spar in steel plants
located at Chicago and points to the west, should tend to sustain a
higher price at mid-continent steel plants than prevails at seaboard.

When all these factors are taken into consideration, it appears
only reasonable to assume that the Colorado producer will find a
fairly ready market at values somewhat lower than those now
obtained, but not as low as formerly quoted. The probability is that
the market will follow in about the same course as that taken by
the metallic minerals, and that the price will settle somewhere between the low prices prevailing during the pre-war period and that
paid during the war, when the demand was far greater than the
supply.

GRAPHITE

INTRODUCTORY: The war much increased the demand for graphite, and the scarcity that resulted was accentuated by the embargo laid by the British and French Governments on the exportation of graphite from Ceylon and Madagascar, while the unsettled conditions in Sonora rendered that source of supply undependable. The net result was the trebling of the price of graphite, and a serious effort to supply the market from local sources, as will be seen from the following tables. Many properties containing amorphous graphite were developed, but no altogether satisfactory deposits of crystalline[1] material were opened:

Source of Supply and Quantity Used Yearly

Imports, pre-war period, 1912-1915..........25,000 short tons
Imports, war period, 1916-1918...............43,000 short tons
Domestic, pre-war period, 1912-1915...... 4,000 short tons
Domestic, war period, 1916-1918.............14,000 short tons

Prices

Per lb.

·Pre-war period, 1912-1915, Crystalline............... 4c to 6c
Pre-war period, 1912-1915, Amorphous..............½c to ¾c
War period, 1916-1918, Crystalline.......................10c to 20c
War period, 1916-1918, Amorphous....................... 1c to 2c

Marketable Quality

For the manufacture of crucibles, graphite should exceed 85% in graphitic carbon and be free from mica, pyrite and iron oxide; a small amount of silica is not objectionable; the product must contain a good percentage of flake or crystalline material. For most other uses it can be of lower grade. For lubricating purposes it must be absolutely free of gritty substances; for pencils the product must be uniform and very pure. Foundry facings absorb large quantities of low-grade material.

There has been a constantly increasing use of the natural graphite in the manufacture of electrodes for use in the electric furnace. The amorphous product is suitable for this purpose. ·

Competition Through Imports

Ceylon, Madagascar, Korea, Italy, Bohemia, Bavaria, Canada and Mexico, all make a production, Ceylon and Madagascar being the most important.

---

[1]The crystalline variety is used in the manufacture of crucibles because its structure is somewhat fibrous and helps to make the crucible tough.
The term "crystalline" applies to those graphites in which the crystal structure is visible to the naked eye. Amorphous graphite has a microscopic crystalline structure.

Occurrence in Colorado

A large vein in Gunnison County, and traceable for more than ten miles from the Tin Cup District to White Pine, is probably the most important deposit of this kind in Colorado. Several hundred tons of very high-grade graphite have been shipped from a few mines located on this vein, the most important being the property near Pitkin, belonging to Woodruff & Woodruff.

State Inspector of Mines, Innes, in his report May 30, 1918, says: "This property is opened by three tunnels driven on the vein, with a winze connecting them; the distance from the lower to the upper tunnel is about 40 feet. Each tunnel is timbered with square sets, and if the lower tunnel had been driven first, and all of the ore stoped out above it, considerable expense would have been saved. Previous to this time, this property produced forty-five, twenty-five-ton cars, for which the owners received $25 per short ton, f. o. b. cars at Quartz Station, which is 5 miles from the mine. Some of this product is said to have been 96% graphitic carbon." At another visit, Inspector Innes describes the ore as averaging between 3 ft. and 4 ft. in thickness.

A well-known authority who visited the property·in 1918, states that an ore of marketable grade, ranging in width from 12 inches to several feet, could be mined. Other occurrences will show much low-grade material carrying Al_2O_3 and SiO_2, which, he says, could be milled to good advantage. The volatile matter ranges from 1% to 18%. This man stated that these properties deserved most careful investigation, for the reason that he found in many places sufficient crystalline material to make the product just as mined suitable for the manufacture of crucibles.

Another competent authority stated that the quantity of high-grade material was so limited that, under normal conditions, this property could not compete with imported graphite, for the reason that it was wholly amorphous, contains some iron, and, except in emergencies, it would not be used by the trade. He was not sufficiently familiar with milling operations to pass judgment upon its amenability. He felt that the freight rates alone would prevent its entering the Eastern markets, to say nothing of the wagon haul and high operating costs, due to its location and the character of the deposit. He said that the product is suitable for the manufacture of electrodes, but the local market for these is so restricted that it would not warrant production of graphite for their manufacture. Scrap material from properties producing good grade crystalline graphite will fix the price of any products used, for which this would be suitable.

Near Turret, in Chaffee County, a similar grade of graphite has been produced. The same kind of material has been found elsewhere in the state, and one of the largest deposits, which carries both flake and amorphous graphite, is in Archuleta County, not far from Pagosa Springs. No work has been done to determine the grade of the ore or the extent of the body. From the information at hand, it appears that it would have to be milled, and even then it may not be possible to bring the grade up to the requirements of the market.

The Gunnison property is shipping at present.

OIL SHALE

By ARTHUR J. HOSKIN

So much has been published in the technical press and in many popular periodicals of our country regarding Colorado's tremeudous resources in oil-yielding shale that the public is mystified by figures cited as pertaining to square miles of such shale beds, and to the average yields of oil per ton. A classical quotation in all such articles is one from Bulletin 641-F of the U. S. Geological Survey, written by Dean Winchester, to the effect that "in Colorado alone there is sufficient shale, in beds that are 3 feet or more thick and capable of yielding more oil than the average shale mined in Scotland, to yield about 20,000,000,000 barrels of crude oil, from which 2,000,000,000 barrels of gasoline may be extracted by ordinary methods of refining. In addition to the oil the same shale should produce with but little added cost about 300,000,000 tons of ammonium sulphate, a compound especially valuable as a fertilizer."

It is not necessary for one to burden his mind with figures of this sort, if he will but accept and remember the one significant fact they bring out, viz., that our state has been late in awakening to the recognition and utilization of one of its most important natural resources, a resource that, according to one authority, will surpass in value of production more than has been or ever will be yielded by all the metal and coal mines of the state. While such a prophecy may be exaggeration, the fact remains that we possess, in our so-called oil shales, the basis upon which will be erected one of America's greatest industries. Colorado does not, by any means, possess such rocks in this country exclusively, but this state does unquestionably contain such rocks in their greatest area, thickness and richness. Oil shales are scattered throughout America, and their exploitation will follow the lead being taken here in Colorado. The discussion following refers exclusively to Colorado affairs :

It was not until 1917 that any real interest was displayed by the public in these shales, which abound in the Counties of Mesa, Garfield, Rio Blanco and Moffat—i. e., along the northwestern border of the state. Although the presence and character of these bituminous sedimentaries had been known to persons residing in that region for many years, no universal attention was directed to them until these formations were investigated and reported upon favorably by experts from a Federal bureau in the bulletin published December 18, 1916, and quoted above. Demand for this pamphlet was such that the edition was soon exhausted. Prospectors were busy that following season, and numerous companies were incorporated. Assuming that success would promptly follow the adoption of Scotch methods of shale treatment, a few small retorts were built in the oil-shale field, especially in the region about DeBeque. It was then discovered that there are differences in oil shales, the world over, and that established foreign processes were futile when applied to the treatment of the bulk of our Colorado shale. Just why this is, we cannot adequately explain. The question is on a par with the one that seeks a reason why, of two lots of coal with seemingly similar characteristics, one will coke and the other will not. Our oil shales exhibit various outward physical properties, usually designated as massive, papery, limey, sandy, asphaltic and waxy. In color they range from black through blue and brown to gray and even to yellow. By far the bulk of our shale is of dark massive varieties that are found, almost invariably, to cake when undergoing distillation in old-style retorts. The first experimental retorts built in the shale field followed well-known Scotch principles. Upon the distillation of a charge of shale, the residue or spent shale was a compact mass, the removal of which entailed much time and effort. The papery shale, which occurs in considerably lesser amount than the massive shale, submits well to distillation in a Scotch-type retort, but prospective operators seem to have arrived at a general conclusion that any retorting plant they install must be capable of treating the various types of shale that occur, for it will be practically impossible to keep these types separate during the operations of mining. For this reason there have come into notice, during the past two years, numerous new schemes for the retorting of our oil shale. Much comment has been elicited from persons who, not knowing the basic reason above explained, have criticised our inventors, as well as honest promoters, who have been impatiently awaiting the demonstration of successful American retorts.

Because of the hindrance just explained, the oil-shale industry showed little progress during 1917. Nor was much constructive work done during 1918, the building of experimental or demonstration units of various types proving slow, with the result that the summer season—when only it is possible to erect plants in the shale field—had passed before satisfactory public demonstrations had been made by inventors. At the same time, some progress was made, and a few companies constructed roads, camps and water-supply pipe-lines, and purchased machinery that was delivered too late for erection last year.

We are at a critical period in this new Colorado industry; the next two years will reveal important advance. New types of retort are being introduced, with every indication of complete success in their proposed purpose. Not a few such types of apparatus originated in Colorado. One or two others have been brought to this state for manufacture and exploitation. It appears certain that Colorado is destined to be the headquarters of this national industry.

It may be in place here to draw attention to the fact that there is no geological relationship between oil shales and oil sands. The popular name of the former is unfortunate, because such shales contain absolutely no, or very little, free oil. Oil sand is a porous rock that is saturated with free petroleum from some source outside itself. Shales that will yield oil when retorted contain original substances that, when heated, generate crude oil and this oil, immediately evaporating, is recovered by condensation outside the retort.

A matter of importance in this industry is the absence of any mineral laws or regulations specifically applying to the location or acquirement of oil-shale lands. Up to the present time, shale claims have been filed upon precisely as though they were petroleum claims, but the point has been repeatedly raised by legal advisers that this is not proper procedure. This land does not fall into any classification of placer ground, in any genetic or practical sense, and it may not be located under the usual placer provisions.

During the greater part of 1918 very few claims were located, for the reason that the oil-land leasing bill, so worded as to cover oil-shale land, was pending before Congress with very strong indications of passage. At present (April, 1919), the status of oil-shale land is so unsettled that no project is contemplated upon land whose location does not date prior to January 1, 1918. It would be well for our citizens, generally, to exercise their influence with the Federal law-makers to the end of expediting whatever legislation is to be enacted in regulating the acquirement of government-owned oil-shale lands.

CEMENT AND PLASTER

Eckel[1] has given a brief review of the Portland Cement resources of Colorado, which includes many analyses of limestones and shales available for the making of cement. Colorado could undoubtedly produce a greater quantity of cement were there a wider market. Two plants operated during the past two years, The Colorado Portland Cement Company, at Portland, and the United States Portland Cement Company, at Concrete. The total production is unknown, but it was probably in the neighborhood of three hundred thousand barrels a year. The price advanced materially since the war broke out.

The gypsum deposits of Colorado are very large, for wherever the "red beds" outcrop, gypsum is pretty sure to be found. Deposits occur all along the Front Range from Wyoming to New Mexico, while on the Western Slope there are enormous outcrops on the Eagle and Roaring Fork Rivers, on Elk Creek, Garfield County, in the Paradox Valley and at Rico. They have been developed at Loveland by the United States Gypsum Company, and by The Colorado Portland Cement Company, at Turkey Creek. All the product goes into Portland cement (as a retarder) or into plaster.

[1] Eckel. Bull. 522, U. S. Geol. Surv., pp. 124-128.

ROCK QUARRIES

The production of building and monumental stone has decreased during the past year, so that most of the granite quarries have closed down, as has also the plant of the Yule Marble Company. The quarrying of limestone for use as flux has increased, however, and the quarries in Fremont and Pueblo Counties have been very active. Lime quarries were operated at Berthoud, Durango, Lime, Monarch, Pueblo, Newett, Thomasville and Turkey Creek.

In view of the fact that there is but little good building stone between the Colorado border and the Mississippi River, an effort should be made to get the building trade of the Middle West to draw more largely on the many excellent stones of Colorado.

The following is a partial list of the available stones of Colorado:

Granite

Granite, suitable for building purposes, is found in almost every mining county in the state, the most famous are the Aberdeen or Gunnison granites (used in the State Capitol), the Salida granite and the Nighthawk granite. The Aberdeen granite, which comes from a quarry on South Beaver Creek, 11 miles from Gunnison, has been described by J. Fred Hunter,[1] as follows:

"The rock is a soda-rich granite, and is known to the trade as gray granite. It approaches a quartz diorite in composition. The thick sprinkling of black biotite through the clear transparent quartz and white, more opaque feldspar, gives the general gray appearance. The rock is entirely crystalline, of medium grain and even texture. The individual crystals average from 2 to 3 millimeters in diameter, few being larger than 7 millimeters.

In thin section the rock is seen to be allotriomorphic— that is, almost all the individual minerals which compose it are without definite crystal outline, and have irregular boundaries. The texture is that usual to granite, and the individual minerals, although varying somewhat in size, show all intermediate gradations, from the smallest to the largest.

The essential mineral constituents, in descending order of abundance are plagioclase of the composition of oligoclase quartz biotite, and potash feldspar (microcline and orthoclase). The potash feldspar is very subordinate, comprising

[1]Bull. 540 K., U. S. Geol. Surv., 1913.

less than 5 per cent. of the rock. The accessory minerals are magnetite, apatite, epidote, calcite and titanite.

An estimate of the mineral percentages of the rock by the ·Rosiwal method gave the following results: Quartz, 36; feldspar, 51; biotite, 12; accessory minerals, 1.

The granite is hard and compact, and is said to work easily. It takes a good polish, becoming slightly darker than on fracture surfaces. It is said to be good for bush-hammer work, and has been used for monumental stone. For the latter purpose, and particularly for inscriptions, the color, susceptibility to polish, and contrast between cut of hammered and ·polished surfaces, are properties of chief economic importance. The abundance of black mineral and the transparency of the quartz and plagioclase feldspar in the granite are very significant in this connection. The granite has a· decided rift (running approximately N. 60° W. in the mass), along which it slabs easily. Physical tests of the granite were made by E. C.' Rhody, a student in the College of Engineering of the University of Colorado, with the following results:

Compressive strength, pounds per square inch..14,340
Modulus of rupture, pounds per square inch...... 2,465
Proper specific gravity 2.71
Apparent specific gravity.................................... 2.70
Ratio of absorption, per cent. .... .17
Porosity, per cent. .... .46
Weight, per cubic foot, pounds ...·...................... 169

In outcrops the granite shows considerable weathering near the surface. In this process it becomes rougher, the quartz and feldspar standing out more prominently, and the rock takes on a brownish and more sombre tone. There are, as a rule, innumerable cracks and minor joints where the rock has been exposed to surface weathering. These, however, are superficial and apparently do not extend more than a short distance from the surface. The face of the quarry shows fresh, unaltered rock, with few joints or cracks. Quarried rock, which has lain out in the weather for several years, shows no signs of staining or disintegration. The granite of the State Capitol at Denver, which came from this quarry, is said to show no evidence of weathering after 20 years' exposure to the weather.''

The Salida Granite is probably the only Colorado granite that has been used extensively for monument work. A description of this stone by R. D. Crawford, Professor of Mineralogy at the University of Colorado, is as follows:

"This dark bluish-gray rock of medium to fine grain chisels well, takes a high polish, and shows lettering on polished faces to good advantage. With one face polished for the inscription, while the other sides are left in the rough, the rock makes a very effective monumental stone. It is sometimes used with all exposed faces polished. For resistance to weathering and permanence of gloss it probably has no superior among granitic rocks.

Though the Salida stone is known commercially as granite, technically it is a quartz-bearing monzonite. It differs from granite in that it contains potash feldspar and soda-lime feldspar in nearly equal quantities, and in that it carries very little quartz. The two feldspars named are bluish gray, and make up about half the volume of the rock. In and among the feldspar grains are a few minute brown crystals of titanite. Nearly half the bulk of the rock is composed of small grains of the black minerals, biotite, hornblende, and magnetite or ilmenite, with a very little pyroxene. The pleasing contrast between light grooves of letters and designs and their dark background of polished surface is due chiefly to the abundance of these black minerals."

The Nighthawk granite is of a very white color, with red streaks, and spangled with black flakes. It dresses well, and is said to take a high polish; because of its white color, however, it will not show chiseled inscriptions. It resists weathering well, and has a tendency to stain. It is used for a building stone. Numeralogically, this granite consists of a fine network of interlocking grains of quartz and a white feldspar. Biotite grains are scattered through, and occasionally grains of magnetite and pyrite are seen. The staining is due to the weathering of these two minerals.

A red granite, called the Silver Plume, has been quarried at Silver Plume, and a gray one known as the Cotopaxi, at Cotopaxi.

A new granite quarry has been opened up near Lyons, but the writer has not seen any of the stone.

Other granites have been quarried locally, and there are undoubtedly many other excellent stones of this type in Colorado.

Igneous Rocks Other Than Granite

At Castle Rock a pink rhyolite tuff has been quarried for building stone, and is satisfactory for small buildings. It has not, however, sufficient strength for use in large buildings. Basalt has been quarried at Trinidad and also at Del Norte for local use, while a dolorite at Valmont has been used to pave streets. Locally, a number of similar stones have been quarried in small quantities for single buildings, while much stone has been quarried for railway ballast and road metal.

Sandstones

The most famous Colorado sandstone is the Lyons, which is a fine-grained stone composed almost wholly of clear quartz grains, but, with enough iron oxide to color it a salmon pink. It is almost quartzitic in hardness, and has a very marked laminated bedding, along which it splits readily. For this reason it has found extensive use for sills, curbs and flagging, but of late years the use of cement for these purposes has almost driven the stone from the market, and it is now used only in buildings. It is well suited for construction by its pleasing color and great durability.

At Turkey Creek also a pink sandstone has been quarried, but is not of such good quality as the Lyons. The Dakota sandstone is hard and quartzitic in character, but locally it contains much pyrite, and quickly stains to a dark brown, and, while it is extraordinarily durable, it is hard to dress. It occurs in almost every county in the state. Many other sandstones have been used to a small extent.

Marble

The most famous deposit of marble in the state is the clear white stone that is quarried by the Yule Marble Company, at Marble, in Gunnison County.

Burchard[1] says:

"The most extensively developed deposits of marble in Colorado are on Yule Creek, in northern Gunnison County. The deposits that are quarried here are high on the left bank of the creek and dip westward at an angle of about 52°. The marble bed is reported to be about 240 feet thick, and to contain four bands of chert, each 2 to 4 feet thick. The underlying rock is cherty blue dolomite, and overlying the marble is a sill of igneous rock which is, in turn, overlain by 500 to 800 feet of blue cherty limestone. The marble itself is, for the most part, white and of medium fine grain, but there are bands of handsome green-stained material within the mass. This quarry has a complete equipment and has maintained a large output of marble for several years. The rock is carried to the mill at Marble, about 3½ miles distant, by an electric tramway. At the marble mill, which is electrically driven and is one of the most completely equipped in the United States, the product is sawed, planed, turned, polished, carved and otherwise prepared for all kinds of interior and exterior construction work. This white marble has been used for all kinds of interior decorations."

---

[1] Mineral Resources of the United States, 1912, Part II, p. 810, U. S. Geol. Survey.

This quarry, which is at present closed, is said to have yielded a marble which in quality was equal to the famous Carrara.

Other marbles have been found in Colorado, for instance, a black marble with much pyrite in it has been found at Boulder, a reddish breccia marble is said to come from the same area, a yellow marble from Canon City, a golden brown onyx marble from Steamboat Springs, and a black and white marble from near Pitkin.

Limestone

Burchard[1] says:

"The limestones of Colorado may conveniently be divided, geographically and geologically, into two groups. The first of these groups includes limestone mostly of Cretaceous age, which occurs in the plains region of the eastern half of the state and in a narrow belt immediately east of the Front Range. The second group includes the limestones mostly of Carboniferous age, which lie west of the Front Range. The two limestone formations of greatest importance in the Cretaceous system are the Niobrara and the Greenhorn. The Niobrara limestone outcrops as a narrow but fairly continuous belt from the Wyoming line southward to Colorado Springs, passing just west of Fort Collins and Denver. South of Colorado Springs are two areas of Niobrara limestone, which occupy much of Pueblo, Otero, Huerfano, Las Animas, Bent, Prowers, Kiowa and Cheyenne Counties, the upper area of outcrop lying along Arkansas River, from near Florence to the Kansas line. The thickness of the Niobrara is about 400 feet, but calcareous shale makes up a considerable part of this thickness."

"In central and western Colorado limestones of Mississippian age cover large areas. Analysis of limestones from a number of points in Garfield, Grand, Gunnison, Jefferson, Park, Pitkin and Summit Counties indicate that this limestone is low in magnesia."

As far as the writer knows, none of this limestone is quarried for any other purposes than for use as a flux or in the manufacture of lime and Portland cement. The limestones in the Niobrara might well be used for building purposes, but the abundance of other good stone prevents this. Limestone is burned for lime in about half of the counties of the state, the large producers are, however, Pueblo, Fremont, Pitkin, La Plata and Chaffee Counties. A little very pure limestone is quarried for sugar factories. The output of stone in Colorado for 1917 and 1918 exceeded $400,000.00 a year.

The following is a directory of Colorado stone quarries:

[1]Op. Cit., p. 812.

DIRECTORY OF STONE QUARRIES IN COLORADO

GRANITE

Boulder County:
  Crescent
  Crags
  Lyons (2)
  Valmont
Chaffee County:
  Granite
  Barre
  Salida (10)
  Turret (3)
Clear Creek County:
  Silver Plume (2)
Douglas County:
  Castle Rock
  Nighthawk
El Paso County:
  Cascade

Fremont County:
  Cotopaxi
  Texas Creek (3)
  Whitehorn
Gunnison County:
  Aberdeen
Jefferson County:
  Buffalo Creek (2)
  Golden
La Plata County:
  Durango
Larimer County:
  Arkins
Pitkin County:
  10 mi. S. E. of Aspen
Rio Grande County:
  Del Norte

SLATE

Gunnison County: Marble (2)

MARBLE

Chaffee County:
  Buena Vista
  Salida
Fremont County:
  Fremont

Gunnison County:
  Marble (2)
Pitkin County:
  Aspen
Saguache County:
  Villa Grove

SANDSTONE

Boulder County:
  Boulder (4)
  Lyons (11) and Noland (5)
Conejos County:
  Osier
Delta County:
  Austin
  Delta (2)

Douglas County:
  Sedalia
Eagle County:
  Peachblow
El Paso County:
  Colorado City (3)
Fremont County:
  Canon City (5)
  Florence

La Plata County:
    Durango (7)
Larimer County: .
    Arkins (4) and Lowerey
    Bellvue and Stout
    Fort Collins (8)
    Loveland
Las Animas County:
    Trinidad (4)

Montrose County:
    Montrose (2)
    Olathe (2)
Pueblo County: .
    Turkey Creek
Rio Grande County:
    Del Norte (2)
Routt County:
    Steamboat Springs.

LIMESTONE AND LIME KILNS

Boulder County:
    Boulder (2)
Chaffee County:
    Newett (3)
Douglas County:
    Platte Canon (silica)
Fremont County:
    Calcite (3)
    Canon City
Gunnison County:
    Los Gunnison (Cement
    • Creek)

La Plata County:
    Rockwood (3)
Larimer County:
    Fort Collins
    Ingleside
Ouray County:
    Ouray
Pitkin County:
    Thomasville (2)
San Miguel County:
    San Miguel

LIMESTONE

Chaffee County:
    Garfield (2)
Douglas County: .
    Near Littleton, Arapahoe
    County
El Paso County:
    Manitou   • ,
Jefferson County:
    Golden .

Morrison
Mesa County:
    Dominguez (near)
Pueblo County:
    Lime
    Livesey
San Juan County:
    Silverton

CLAY PRODUCTS

The production of the clay products used in building has decreased during the past two years and the number of men employed was also smaller. On the other hand, the production of high-grade porcelain for use in chemical analysis has much increased, as has also that of refractory brick and of tile. This is, of course, a reflection of conditions in the building trade. The most noteworthy development has been the expansion of the Herold China and Pottery Company (at Golden), until a large part of the chemical porcelain used in the United States is manufactured at their plant. The ware is better than the Royal Berlin Porcelain, which was the standard until the war broke out.

The clay mines at Golden have been active, and a large tonnage of fire clay has been mined. Local companies have made building brick at several towns, while there has been a lively demand for silo tile, particularly of an alkali-resistant type. These are rather hard to make, as the kiln operator must fire them just enough to vitrify the surface, and yet not so vigorously that the tile will lose its shape. There have been many inquiries for pottery clays and kaolin.

The resumption of building activities is expected to bring renewed prosperity to the other branches of the clay-working trade.

GEMS IN COLORADO

Small quantities of semi-precious stones are mined in Colorado every year. From Crystal Peak, in Teller County, Devil's Head, in Douglas County, and Crystal Park, El Paso County, Amazonstone (green microcline), smoky quartz (cairngorm stone, smoky topaz), rock crystal, orthoclase, topaz and phenacite are mined and sold in the curio stores of Denver and Colorado Springs. Amethyst has been mined at St. Peter's Dome, in El Paso County, tourmaline at Royal Gorge, in Fremont County, and aquamarines (light blue beryl) at Mount Antero. Jasper, chalcedony and agate are found in the gravels of the Tertiary formations that cover the great plains, and agate is also abundant in northwestern Colorado. Small selvedges of turquoise of fair color have been found in Conejos County, while garnets, rose quartz, hematite, pyrite and rhodonite have also been mined. None of these stones can be called more than semi-precious, and their chief value lies in the cutting which they receive. It is probable, however, that they will continue for many years to add from five to ten thousand dollars a year to the mineral production of Colorado.

MINOR NON-METALLIC PRODUCTS

Fuller's Earth

A deposit of good grade has been opened at Canon City.

Mineral Waters

The shipment of bottled Manitou water has continued, and this product is said to be increasing in popularity.

Barytes

Barytes was mined in Boulder and Custer Counties. The amount, however, was small. The Boulder deposit is said to be objectionable on account of the presence of iron oxide and quartz.

Lithium

Some natro-amblygonite was mined at Royal Gorge several years ago.

Abrasives

Grindstone and buhrstone have been mined for local use north of Lyons and in Gunnison County. Garnet, in sufficient quantity to be used as an abrasive, occurs in Chaffee County, and corundum is also reported from that locality.

Silica

Certain sandstones in Fremont County are quarried for flux. However, the amount quarried each year is small. There is a demand for a good molders' sand, which has not as yet been supplied from this state. For this work a sand should be free from all impurities, except possibly some iron oxide; the grains should be rounded and rather coarse. Locally, much sand has been mined for building purposes.

AUTHORITIES FOR TOTAL PRODUCTION OF STATE

By C. W. Henderson, Statistician, U. S. Geological Survey.

In the table, "Total Production of Colorado, 1858-1915, by Years," as many sources of information as were available were used, but primarily the sources were, for gold and silver: From 1868-1875, Raymond's reports on statistics of mines and mining in the states and territories west of the Rocky Mountains; from 1876-1879, reports of the Director of the Mint, showing Colorado total only; from 1879-1896, the Director of the Mint's final figures for the state; from 1897-1904, reports of the Colorado State Bureau of Mines (found to check very well with the Director of the Mint reports, and also in better form by counties, particularly for copper and lead); from 1905-1915, U. S. Geological Survey Mineral Resources (Mines Report). For copper: Raymond's reports were used for the early years, but much copper by estimation from other sources, not hitherto credited to Colorado, was given the state; up to 1896, inclusive, Kirchhoff's table of value of copper product from the beginning of mining operations in Colorado to 1882, in the general copper report, U. S. Geological Survey Mineral Resources for 1881-2, p. 228, was also drawn upon, as was Butler's general copper report, showing smelter production of copper in Colorado from 1874-1910, in U. S. Geological Survey Mineral Resources, 1910, part 1, pp. 171-173; from 1897-1904, the smelter ore receipts, as shown by the Colorado State Bureau of Mines, were used; and from 1905-1915, the mines report of the U. S. Geological Survey Mineral Resources were used. For lead: in the early days Raymond's reports were the principal source, but, as he paid little attention to the total content of lead shipped or produced, a great deal of estimating has been done from Raymond's data of tonnages and grade shipped, thus crediting, for the first time, to Georgetown, Clear Creek County, in particular—and to other counties—a large quantity of lead in ores shipped to Eastern smelters in the United States and to Wales and Germany; much data has been drawn also from Kirchhoff's general lead report, showing production of lead in Colorado from 1873-1882, in U. S. Geological Survey Mineral Resources for 1881-2, p. 310, but for the early days this table appears to be only the output of lead from lead smelters in Colorado, and not including lead in ores shipped; from 1882-1896, the figures have been taken from Kirchhoff's annual general lead reports of the U. S. Geological Survey Mineral Resources; from 1897-1904, the Colorado State Bureau of Mines reports; and from 1905-1915, the mines report of Colorado, in U. S. Geological Survey Mineral Resources. For zinc: the Colorado State Bureau of Mines reports first show a production of zinc in 1902, and for the years 1902 to 1907, these figures are used, and from 1908-1915, the mines report of the U. S. Geological Survey,

Mineral Resources. Zinc production in Colorado, however, began in 1885, when low-grade zinc ore was treated in the East until 1891, when F. L. Bartlett erected the zinc oxide plant at Canon City; from 1891-1897 this was the only plant recovering zinc from Colorado ores, but in 1898 a small quantity of Colorado ores found a market in the United States, and in 1899 found a market by water to Belgium, and soon after, spelter plants in the United States found they could handle the ferruginous zinc blende ores of Colorado. For the years 1885-1891, therefore, an estimate has been made on the verbal statement of Mr. Bartlett; from 1891-1901, inclusive, much of the data has been drawn from annual review editions of the Denver Republican and the Leadville Herald-Democrat, and from annual volumes of Mineral Industry.

Preston[1] says in regard to the authenticity of Raymond's report:

"Complete and accurate statistical data were furnished by a very large number of mining companies, but it is easy to recognize in his report Mr. Raymond's reluctance to give numerical data of the total production of the precious metals as the result of his own special investigations, although the public was most anxious for such data. But this very reluctance of the author to make estimates of the total production because of the insufficiency of the material, inspires all the greater confidence in his own special data. When Mr. Raymond, as he occasionally does, but with the utmost reservation, makes estimates of totals, they rightly claim authority over all others, so long as there is no definite reason given for material deviation from them."

Preston[2] says in regard to reports entitled "Report of the Director of the Mint on the Production of the Precious Metals," first published in 1881 (for 1880), by Mr. Burchard:

"The total production of gold and silver in the United States and the probable yield of the mines of each state and territory have been annually estimated by the Director of the Mint and published in his reports. These statements have been based upon information obtained from officers of the mints and others in the mining regions, and upon statements of depositors as to the locality of production of the gold and silver received at the various mints and assay offices. From these incomplete data approximate estimates have been annually made, which, although probably inaccurate in many details, have been found exceedingly useful for statistical purposes.

"The Mint Bureau, through its subordinate institutions, possesses unusually excellent facilities for obtaining such information, as *during the last fiscal year nearly all the gold produced by the mines of the country was deposited at the mints and assay offices*

[1]Op. Cit., p. 26.
[2]Op. Cit., p. 39.

and exchanged for coin or bars, and of the silver produced in the United States more than three-fourths was purchased by the Government for coinage or deposited for bars. The depositor or seller himself furnished the locality of production of nearly half of the gold and about one-third of the silver deposited or purchased, the balance being from unknown localities.

"To ascertain the state or territory in which the latter was produced, it became necessary to employ other means and look elsewhere for information. This information has been sought, both by letter or personal interview, from mine and mill owners, smelting, refining, and reduction works, banks and bullion brokers, express companies, railroad and freight agents, and customhouses. The information thus obtained, while incomplete in detail from any one source, has been of great value as a means of comparison and arriving at general results. Six refining and reduction works alone, east of the Rocky Mountains, treated more than half of the silver product of the country.

"It was deemed advisable to assign to mint officers, or other competent persons in the mining regions, the territory in their immediate vicinity, with instructions to procure from the officers, agents, and owners of mines, mills and reduction works as full and detailed information as possible, while the statistics of the amount of gold and silver ore and bullion treated at or transported to other refineries, and placed upon the market and exported, was left to be ascertained under the immediate direction of this office."

And in Burchard's last report on the production of the precious metals in the United States, he says:

"While it might have sufficed to have published merely the results of these investigations embodied in the estimates of the total product of gold and silver during the year, it seemed desirable to trace the bullion known to have been converted into coin, or manufactured into bars and exported, or used in the arts, back to the district or mining region from which it was derived.

"This work necessitated the accumulation of statistics as full as possible of the treatment, movement, and final disposition of the bullion from its extraction from the earth in the form of ore or native metal to its delivery to the Government for coinage or to private parties for export or other use. Supplementing this with direct information as to the actual yield of the most important mines, and with careful and generally reliable reports from correspondents and mint officers in regard to the condition and progress during the year of mining for the precious metals in the principal localities, states and territories producing gold and silver, it is believed that sufficient data has been obtained to justify an estimation of the yield of the various counties, as well as of the states and territories.

"It would be a gratification to me to publish all the detailed information in regard to the actual product of individual mines, so that the accuracy of my conclusions could be tested by the data upon which they are based, but most of these facts were confidentially disclosed and not to be made public, except with express permission."

Horatio C. Burchard was succeeded as Director of the Mint by Dr. James P. Kimball in 1885. Dr. Kimball's method of estimating the production of gold and silver in the United States may be inferred from the two tables in the report of the Director of the Mint upon the production of the precious metals in the United States during the calendar year 1885, pages 18 and 23, respectively. * * *

Since 1890, when Mr. E. O. Leech, the successor of Dr. James P. Kimball as Director of the Mint, published his first report on the Production of Gold and Silver in the United States, this Bureau has made each year two independent calculations of the gold and silver product of the United States in the year under consideration and taken their average as the actual product. A clear idea of the way in which these calculations are made at present may be formed by consulting the introductory pages of the present report.

The most competent statisticians have approved and commended the method now employed by the Bureau. The late Dr. Soetbeer, undoubtedly the most competent judge in such matters in the present or any other generation, said of it, in Hildebrand's Jahrbucher, 1891, pages 523, 546, and 547:

"To show by an example in what manner the Bureau of the Mint has endeavored to ascertain, as nearly as possible, the precious metal production of the United States, we may, by way of explanation, state how the amount of the gold product of the country for the year 1889 was reached.

"With the exception of the comparatively small amounts of gold, newly produced from domestic mines, which are immediately employed by goldsmiths, all the gold product of the country finds its way to public institutions, either in the form of unrefined bullion directly from the mines or gold washings, or in the form of fine gold bars from private parting establishments. The charges for refining at the Mints, as well as for the attestation by them of the weight and fineness of the bars deposited there, are insignificant. Coinage is done gratis for the owner, and this is true also of the transportation to the Mint at Philadelphia of the gold to be coined.

"In this way the total gold product of the country, so to say, comes to the knowledge of the Mint Bureau. During the year 1889, 689,658 ounces of gold in the form of unrefined bullion were

deposited at the mints; and of the 911,676 ounces of fine gold refined at private institutions, 856,361 ounces were stamped at the
assay office in New York and at the mint in San Francisco. There
are also special data relating to the quantity of gold imported from
foreign countries, especially from Mexico and British Columbia,
into the United States, and directly employed or refined there.''

The statistics of the production of gold in the year 1889 were,
therefore, made up as follows:

SOURCES

	Fine Ounces
Bullion of domestic production deposited at mints and assay offices	1,546,019
Bullion of domestic production (other than United States Mint or assay office bars) exported from the United States	54,012
Bullion of domestic production reported by 26 private refineries in the United States as having been made into bars for manufacturers and jewelers	50,009
Total	1,650,040

Deduct:

Foreign bullion reported by private refineries in the United States as contained in their product of fine gold bars deposited at mints and assay offices, and there classified as of domestic production	63,811
Domestic product for 1889	1,586,229

A further and independent calculation, according to data
from the several mining districts, gave for the year 1889 the domestic product of gold at 1,587,632 ounces, $32,817,190. The
average of these two results, $32,800,000 in round numbers, or
1,587,000 ounces, has been officially recognized as the product of
gold in the United States during the year 1889.

From this example we can see from what material and by
what method the statistics of the production of the precious metals
in the United States are obtained. They are the result of careful,
thorough investigation, and not a superficial estimate. * * *

The method at present followed by the Bureau, to ascertain
the annual output of the gold and silver mines of the United States,
reduces to a minimum the possibilities of serious error in the calculation of the amount of that output. It is scarcely possible that
the results reached by it can vary to any great extent from the
actual yield of the gold and silver mines of the country in the
year under consideration. The object to be realized being to

obtain the amount of the finished product of gold and silver put on the market in a given year, the Bureau has to depend only on the Government's records of its own mints and assay offices, the records of its exports, and information furnished by the owners of private refineries. But for the courtesy of the latter in furnishing the amounts of the precious metals prepared for the market in their establishments the difficulties in the way of reaching reliable figures in this domain would be as great as those that confronted the early statisticians, who concerned themselves with the annual production of gold and silver in the United States.

No higher praise can be bestowed on the calculations made by the Bureau of the Mint of the production of the precious metals in the United States; no stronger evidence of their accuracy and reliability can be produced than that volunteered by the late Dr. Soetbeer in the words: "They are result of careful, thorough investigation, and not a superficial estimate."

Indeed, the United States is the only great producing country whose gold and silver output from its own mines is at the present day ascertained with practical certainty, and this certainty is attained by the method it pursues in determining the amount of the finished product of the precious metals it puts annually on the market.

The gold product of Australasia is doubtless approximately correct, but only approximately so, because there is no certainty that the crude gold reported as produced in the colonies averages .920 fine. The same may be said of the gold yield of the·South African Republic, whose output of crude gold is uniformly estimated to be .847½ fine.

The yearly product of Russia is taken to be equal to the amount deposited at the St. Petersburg mint, and does not represent the total yield, a large quantity being clandestinely exported through China and escaping declaration in other ways.

Germany gives the total amount of gold and silver refined in the country, but the information relative to the output of its own mines is not altogether satisfactory.

It would be difficult, if not impossible, however, to devise a method by which that of the United States could be more accurately determined than it is by the process employed for some years past by the Bureau of the Mint.

The Colorado State Bureau of Mines reports have been found to be very accurate, with some possible errors in origin by counties, an error passed to the Bureau, due to shipping point of mine not always being in same county as the mine.

For' the mines report of the U. S. Geological Survey Mineral Resources, the reader is referred to the method of collecting statistics by H. D. McCaskey, in Mineral Resources for 1914, pt. 1, pp. 835-836, in which he has compared the results of the mint and mine returns for the 10-year period, 1905-1914.

The foot-notes to the table of total production of Colorado for 1859-1915, show in more detail the exact source of the figures.

SHOWING BY COUNTIES THE TOTAL MINERAL PRODUCTION OF COLORADO TO 1917.

Name of County.	GOLD Value	SILVER Fine Ounces	SILVER Value	LEAD Pounds	LEAD Value	COPPER Pounds	COPPER Value	ZINC Pounds	ZINC Value	Total
Arapahoe, 1858-1904	$ 8,101	101	$ 64	....	$	....	$	....	$	$ 8,465
Archuleta, 1897-1904	1,489	505	302	....	....	....	....	....	....	1,791
Baca, 1906-1917	....	285	164	286,522	12,241	4,411	140,751	....	....	22,365,794
Boulder, 1868-1917	16,707,428	7,190,137	6,724,646	6,618,623	286,522	931,516	140,751	....	....	20,030,280
Chaffee, 1859-1917	7,139,364	4,396,355	3,994,188	155,889,456	5,451,453	9,172,675	2,364,871	2,138,314	113,878	20,837,793
Clear Creek, 1859-1917	22,969,474	56,349,314	50,583,492	168,953,311	7,362,312	11,273,316	1,791,457	1,916,695	24,764	84,057,791
Conejos, 1889-1917	82,612	52,715	31,524	58,048	1,802	237	217	....	....	72,659
Costilla, 1875-1916	43,416	4,272	2,552	202	....	....	....	....	....	7,588,632
Custer, 1872-1917	2,173,211	4,239,511	4,134,885	20,300,274	1,032,424	331,173	64,345	18,567	1,314	7,588,632
Delta, 1894-1915	4,273	206	176	....	....	....	....	....	....	4,449
Dolores, 1879-1917	1,962,949	11,468,112	8,956,793	35,465,224	1,648,405	9,272,141	3,931,828	9,690,171	636,640	11,073,104
Douglas, 1875-1915	4,273	....	....	....	....	....	....	....	....	....
Eagle, 1879-1915	2,145,464	5,461,263	4,399,538	83,268,163	3,602,817	2,059,043	931,828	9,463,164	636,640	20,534,151
El Paso, 1913-1914	....	....	....	....	....	13,276	2,000	....	....	2,000
Fremont, 1881-1917	80,778	99,315	84,484	677,600	28,400	644,429	114,832	194,332	12,877	413,878
Garfield, 1885-1917	18,387	....	....	....	....	162	164	....	....	18,613
Gilpin, 1859-1917	83,411,327	10,177,088	8,197,678	23,186,451	1,406,386	24,565,114	3,988,677	301,614	24,764	97,023,820
Grand, 1889-1916; 1917	2,113,182	6,345,195	4,916,924	39,663,357	1,723,972	95,171	168,805	3,787,843	960,087	9,260,311
Gunnison, 1861; 1872-1917	2,113,529	6,346,714	4,918,924	....	....	935,669	....	....	....	....
Hinsdale, 1871-1917	1,423,474	4,925,698	56,172,163	3,926,091	2,801,721	392,338	1,164,034	57,528	10,336,921	....
Huerfano, 1873; 1888-1907	62,296	12,465,681	10,863	3,195,224	258	19,656	3,160	....	....	170,416
Jefferson, 1858; 1886-1916	47,588,874	221,883,019	181,666,622	81,717,438	11,461	94,269,485	13,279,567	401,332,828	16,794,157	401,332,828
Lake, 1860-1917	4,479,942	1,683,564	1,960,083	249,952	11,461	44,707	13,270,567	16,794,157	....	4,587,184
La Plata and Montezuma, 1873-1917	....	....	....	1,868	11,461	376,485	1,126,016,828	....	....	....
Larimer and Jackson, 1895-1917	21,304	2,562	1,715	....	....	235,328	38,647	30,712	1,659	66,345
Las Animas, 1866; 1887-1899	2,094	29	15	....	....	26,598	5,222	....	....	2,109
Mesa, 1883-1915	6,423	9,934	5,970	8,501,171	8,561,171	263,714	42,186	27,476,123	1,664,916	13,131
Mineral, 1889-1917	2,686,377	42,757,945	27,355,810	194,974,241	....	453,616	82,023	....	....	49,147,468
Montrose, 1886-1917	46,259	184,448	118,724	64	....	....	....	....	....	250,019
Ouray, 1873-1917	34,675,607	37,843,100	27,588,110	151,703,744	6,482,505	22,387,879	3,207,340	1,138,910	72,048,698	72,048,698
Park, 1859-1917	9,735,098	9,732,817	6,679,041	38,726,677	1,672,110	1,275,546	370,991	9,973,552	189,512	18,675,591
Pitkin, 1883-1917	972,753	94,233,623	535,164,109	535,164,109	21,892,893	1,118,546	15,256,526	15,256,526	948,037	95,523,581
Pueblo, 1894; 1896; 1900; 1901	179,158	90	55	46,802	3,056	210	35	....	....	2,557,828
Rio Grande, 1870-1917	2,363,166	179,158	112,691	46,802	3,056	124,111	19,916	....	....	2,557,828
Routt and Moffat, 1866; 1873-1917	384,629	1,324,386	1,106,727	7,398,839	336,224	778,529	16,794	69,081	....	431,979
Saguache, 1880-1917	21,778,759	26,534,695	18,299,546	271,798,239	12,215,249	649,480	121,124	1,873,522	1,863,318	61,215,964
San Juan, 1873-1917	49,464,401	34,353,696	23,809,863	129,387,902	6,103,285	45,136,915	6,909,418	24,433,213	1,139,375	83,603,851
San Miguel, 1875-1917	16,898,911	12,895,696	11,027,402	147,047,516	6,376,610	16,631,668	1,135,623	109,751,327	8,270,083	42,790,419
Summit, 1860-1917	....	....	....	....	....	995,689	128,829	109,751,327	8,270,083	294,142,092
Teller, 1891-1917	293,202,811	1,558,596	933,258	....	....	....	....	....	....	294,142,092
Miscellaneous, small counties, 1888...	8,785	1,214	1,111	....	....	....	....	....	....	9,926
Totals	**$623,047,160**	**693,796,442**	**$466,463,217**	**3,962,148,896**	**$173,905,020**	**237,422,283**	**$ 35,756,138**	**1,484,929,849**	**$106,210,930**	**$1,405,684,255**

Compiled from figures by C. W. Henderson, Statistician, U. S. Geological Survey.

TOTAL MINERAL PRODUCTION BY YEARS FOR ARCHULETA COUNTY

Years	GOLD Value	SILVER Fine Ounces	SILVER Value	LEAD Pounds	LEAD Value	COPPER Pounds	COPPER Value	ZINC Pounds	ZINC Value	Total
1897	$703	348	$209							$ 912
1898	145	40	24							169
1899	103	43	26							129
1900	145	30	18							163
1901	124	18	11							135
1902	83	10	5							88
1903	62	6	3							65
1904	124	10	6							130
Grand total..	$1,489	505	$302							$1,791
1918	No									

No production since 1904.
Compiled by figures from C. W. Henderson, Statistician, U. S. Geological Survey, and from other sources.

TOTAL MINERAL PRODUCTION BY YEARS FOR ARAPAHOE COUNTY

Years	GOLD Value	SILVER Fine Ounces	SILVER Value	LEAD Pounds	LEAD Value	COPPER Pounds	COPPER Value	ZINC Pounds	ZINC Value	Total
1885	$271									$ 271
1886	293									293
1887	177									177
1888										
1889										
1890										
1891										
1892										
1893	86									86
1894	1,081	59	$38							1,119
1895	1,894	19	13							1,907
1896	2,108	14	8							2,116
1897	703	7	4							704
1898	269	2	1							270
1899	248									248
1900	331									331
1901	227									227
1902	165									165
1903	248									248
1904										
Grand total..	$8,101	101	$64							$8,165
1918	No production.									

Compiled from figures by C. W. Henderson, Statistician, U. S. Geological Survey, and from other sources.

TOTAL MINERAL PRODUCTION BY YEARS FOR BACA COUNTY

Years	GOLD Value	SILVER Fine Ounces	SILVER Value	LEAD Pounds	LEAD Value	COPPER Pounds	COPPER Value	ZINC Pounds	ZINC Value	Total
1900	$103	102	$ 63			8,900	$1,477			$1,643
1901	83	80	48			8,590	99			230
1902	103	59	31			1,928	235			369
1903-1914, no production.										
1915		8	4			504	90			94
1916		50	33			2,772	682			715
1917	3	57	47			6,806	1,858			1,908
Total	$292	356	$226			21,511	$4,441			$4,959

1918, no production.

Compiled from figures by C. W. Henderson, Statistician, U. S. Geological Survey, and from other sources.

TOTAL MINERAL PRODUCTION BY YEARS FOR BOULDER COUNTY

Years	GOLD Value	SILVER Fine Ounces	SILVER Value	LEAD Pounds	LEAD Value	COPPER Pounds	COPPER Value	ZINC Pounds	ZINC Value	Total
1858-1862	$ 100,000									$ 100,000
1863	25,000									25,000
1864	25,000									25,000
1865	20,000									20,000
1866	15,000									15,000
1867	10,000									10,000
1868	50,000									50,000
1869	100,000	3,547	$ 4,700							104,700
1870	100,000	60,241	80,000							180,000
1871	156,605	60,377	80,000							236,605
1872	224,852	199,414	263,625							488,477
1873	155,000	282,326	366,177							521,177
1874	160,000	293,806	375,484							535,484
1875	218,086	203,344	252,147							470,233
1876	200,000	232,031	269,156							469,156
1877	400,000	232,031	278,437							678,437
1878	40,000	270,703	311,308							711,308
1879	400,000	348,047	389,813							789,813
1880	300,000	425,391	489,200							789,200
1881	200,000	270,703	305,894							505,894

Year										
1882	260,000	239,766	273,333							533,333
1883	300,000	123,750	137,363							437,363
1884	350,000	100,547	111,607							461,607
1885	300,000	84,691	90,619							390,619
1886	382,185	84,691	83,844							466,029
1887	253,546	70,091	68,689	593	27					322,262
1888	189,241	230,205	216,393	246,282	10,836					416,470
1889	344,503	174,471	164,003	51,215	1,997	2,748	371			510,874
1890	380,059	118,898	124,843	45,894	2,065	90,691	14,148			521,115
1891	683,941	1,690	41,273							725,214
1892	982,988	182,156	158,476	9,697	388					1,141,852
1893	479,665	257,462	200,820	10,000	370	50,000	5,400			686,255
1894	489,592	75,730	47,710	10,000	330	50,000	4,750			542,382
1895	401,926	40,685	26,445	11,439	366	57,864	6,191			434,928
1896	385,653	79,047	53,752	4,216	126	63,252	6,831			446,362
1897	512,657	138,715	83,229	309,115	11,128	58,474	7,017			614,031
1898	581,302	91,432	53,945	8,967	341	22,452	2,784			638,372
1899	547,858	76,371	45,823	28,043	1,262	78,816	13,478			608,421
1900	607,016	90,327	56,003	76,076	3,347	20,371	3,382			669,748
1901	774,298	113,782	68,269	191,987	8,255	22,186	3,705			854,527
1902	538,702	82,710	43,836	13,493	553	11,090	1,353			584,444
1903	431,569	61,833	33,390	115,100	4,834	6,154	843			470,636
1904	411,581	57,424	33,306	62,111	2,671	26,115	3,343			450,901
1905	261,601	70,921	43,262			2,227	347			305,210
1906	188,769	21,923	14,908	47,491	2,707	3,539	683			207,067
1907	161,658	23,480	15,497	16,491	874	22,840	4,568			182,597
1908	147,234	21,498	11,394	96,503	4,053	28,955	3,822			166,503
1909	163,273	48,183	25,055	425,605	18,301	16,485	2,143			208,772
1910	139,911	46,517	25,119	53,205	2,343	16,772	2,110			169,503
1911	163,174	53,753	28,489	145,955	6,568	27,752	3,469			201,700
1912	119,426	72,335	44,486	305,822	13,762	22,176	3,659			181,333
1913	69,274	162,384	98,080	409,500	18,018	25,535	3,958			189,330
1914	93,710	312,217	172,656	523,821	20,429	24,316	3,234			295,029
1915	160,433	271,292	137,645	890,042	41,832	86,680	15,169			354,979
1916	119,299	292,824	192,678	864,333	59,639	64,707	15,918			387,534
1917	66,841	294,375	242,565	575,582	49,500	29,513	8,057			366,963
Totals	$15,707,428	7,190,137	$6,734,646	5,548,623	$286,922	931,710	$140,753			$22,869,749
*1918	$55,000	159,282	$154,141	284,300	$20,499	1,500	$4,317			$233,957

*Estimate.

Compiled from figures by C. W. Henderson, Statistician, U. S. Geological Survey, and from other sources.

TOTAL MINERAL PRODUCTION BY YEARS FOR CHAFFEE COUNTY

Years	GOLD Value	SILVER Fine Ounces	SILVER Value	LEAD Pounds	LEAD Value	COPPER Pounds	COPPER Value	ZINC Pounds	ZINC Value	Total
1859-67	$380,000									$380,000
1868	10,000									10,000
1869	10,000									10,000
1870	70,000									70,000
1871	10,000									10,000
1872	10,000									10,000
1873	10,000									10,000
1874	10,000									10,000
1875	21,551									21,551
1876	25,000	3,867	$4,486							29,486
1877	25,000	7,734	9,281	50,000	$2,750					37,031
1878	25,000	7,734	8,894	50,000	1,800					35,694
1879	28,500	30,938	34,651	50,000	2,050					65,201
1880	31,500	61,875	71,156	100,000	5,000					107,656
1881	50,000	127,617	144,207	500,000	24,000					218,207
1882	45,000	77,344	88,172	1,000,000	49,000					182,172
1883	50,000	204,961	227,507	4,300,000	184,900					462,407
1884	80,000	146,953	163,118	12,000,000	444,000					687,118
1885	100,000	200,000	214,000	18,700,000	729,300					1,043,300
1886	313,917	332,965	329,635	13,000,000	598,000					1,241,552
1887	409,050	423,738	415,263	14,954,155	672,937					1,947,250
1888	393,457	292,349	274,808	8,743,053	384,694					1,052,959
1889	299,853	137,759	129,493	5,000,000	195,000					624,346
1890	254,250	145,674	152,958	2,400,000	108,000					515,208
1891	279,060	64,830	64,182	1,100,000	47,300					390,542
1892	147,203	85,632	74,500	6,324,319	252,973			100,000	$4,600	479,276
1893	154,164	92,448	72,109	4,000,000	148,000	50,000	$5,400	100,000	4,600	383,673
1894	120,565	25,527	16,082	1,100,000	36,300	50,000	4,750	100,000	3,500	181,197
1895	153,629	29,630	19,260	285,056	9,122	76,070	8,1 0	120,000	4,320	194,471
1896	193,465	151,738	103,182	1,047,310	31,419	559	60	120,000	4,680	332,806
1897	226,936	53,859	32,315	196,301	60,710	172,891	20,747	100,000	4,100	344,808
1898	227,535	85,273	50,311	2,522,554	95,857	114,202	14,161	100,000	4,600	392,464
1899	216,663	147,339	88,403	1,193,70	53,688	693,736	119,142	100,000	5,800	483,696
1900	172,677	125,330	7,705	833,42	36,672	753,677	125,110	100,000	4,400	416,564
1901	158,684	76,286	45,772	209,788	9,020	576,251	96,234	100,000	4,100	313,810
1902	417,513	114,155	60,502	456,889	18,732	173,538	21,172	220,500	10,584	528,503

TOTAL MINERAL PRODUCTION BY YEARS FOR CLEAR CREEK COUNTY

Years	GOLD Value	SILVER Fine Ounces	SILVER Value	LEAD Pounds	LEAD Value	COPPER Pounds	COPPER Value	ZINC Pounds	ZINC Value	Total
1859-65	$2,000,000	....	$	....	$	....	$	....	....	$ 2,000,000
1866	50,000	15,123	20,250	....	....	....	....	....	....	70,250
1867	50,000	15,220	20,251	....	....	....	....	....	....	70,251
1868	50,000	106,953	141,820	....	....	....	....	....	....	191,820
1869	50,000	377,359	500,000	100,000	6,000	2,000	485	....	....	556,485
1870	80,000	362,465	481,354	200,000	12,000	2,500	530	....	....	573,884
1871	20,000	640,790	849,047	550,000	33,000	3,000	724	....	....	902,771
1872	25,000	1,118,299	1,478,391	1,000,000	64,000	4,000	1,422	....	....	1,568,813
1873	34,000	902,668	1,170,760	1,000,000	60,000	10,000	2,800	....	....	1,267,560
1874	42,500	1,634,434	2,088,807	803,983	48,239	15,000	3,300	....	....	2,182,846
1875	72,408	1,343,610	1,666,076	1,300,000	75,400	15,000	3,405	....	....	1,817,289
1876	95,161	1,421,104	1,648,481	819,672	50,000	15,000	3,150	....	....	1,796,792
1877	96,500	1,534,560	1,841,472	2,236,364	123,000	15,000	250	....	....	2,063,822
1878	134,000	1,759,652	2,023,600	2,722,222	98,000	25,000	4,150	....	....	2,259,750
1879	120,000	1,546,875	1,732,500	1,951,219	80,000	100,000	18,600	....	....	1,951,100
1903	169,329	129,900	70,146	249,308	10,471	79,581	10,903	3,000	162	261,011
1904	64,346	69,045	40,046	652,238	28,046	263,239	33,695	294,440	15,016	181,149
1905	2,378	75,265	45,912	1,250,302	58,764	869,507	135,643	849,963	50,148	322,845
1906	156	54,609	37,134	1,227,019	69,940	349,466	67,447	623,855	38,061	271,518
1907	75,364	38,465	25,387	630,623	33,423	799,505	159,901	2,407,730	142,056	436,131
1908	49,057	35,745	18,945	1,040,238	43,690	337,804	44,590	703,706	33,074	189,356
1909	30,485	35,477	18,448	584,492	25,133	568,868	73,953	947,741	51,178	199,197
1910	77,152	182,003	98,282	970,523	42,703	226,772	28,800	438,539	23,681	270,618
1911	5,514	92,098	48,812	1,001,651	45,074	88,448	11,056	200,509	11,429	182,085
1912	97,489	104,686	64,382	992,578	44,666	133,570	22,039	736,392	50,811	279,387
1913	312,892	168,985	102,067	3,196,545	140,648	315,011	48,827	2,121,977	118,829	723,263
1914	332,230	272,242	150,550	3,690,359	143,924	319,496	42,493	2,173,177	110,832	780,029
1915	316,146	226,996	115,087	3,630,127	170,616	384,046	60,908	4,676,355	579,868	1,242,625
1916	185,050	100,749	66,293	3,016,999	208,166	1,001,455	246,358	4,744,985	635,828	1,341,695
1917	133,624	146,535	120,745	2,150,523	184,945	807,883	220,552	2,181,932	222,557	882,423
Totals	$7,130,364	4,986,355	$3,994,188	125,889,456	$5,451,433	9,172,575	$1,622,181	24,364,871	$2,138,214	$20,336,280
*1918	$110,000	62,580	$60,570	1,871,586	$140,005	247,040	$60,698	2,385,790	$188,600	$559,873

*Estimate.
Compiled from figures by C. W. Henderson, Statistician, U. S. Geological Survey, and from other sources.

TOTAL MINERAL PRODUCTION BY YEARS FOR CLEAR CREEK COUNTY—Continued

Years	GOLD Value	SILVER Fine Ounces	SILVER Value	LEAD Pounds	LEAD Value	COPPER Pounds	COPPER Value	ZINC Pounds	ZINC Value	Total
1880	196,000	1,902,650	2,188,054	1,517,500	25,875	200,000	42,800	....	....	2,452,729
1881	200,000	1,546,875	1,747,969	815,000	39,120	200,000	36,400	....	....	2,023,489
1882	220,000	1,299,375	1,481,288	815,000	39,935	200,000	57,300	....	....	1,798,523
1883	250,000	1,222,031	1,356,454	815,000	35,045	300,000	49,500	....	....	1,690,999
1884	600,000	1,314,844	1,459,477	1,038,273	38,416	300,000	39,000	....	....	2,136,893
1885	500,000	1,356,364	1,450,920	1,038,273	40,493	200,000	21,600	25,000	1,075	2,014,088
1886	609,070	1,396,364	1,382,400	1,630,000	74,980	200,000	22,200	25,000	1,100	2,089,750
1887	317,214	1,284,083	1,258,401	1,847,930	83,157	200,000	27,600	25,000	1,150	1,687,522
1888	419,821	1,148,190	1,079,299	3,761,246	165,495	200,000	33,600	75,000	3,675	1,701,890
1889	521,909	1,770,875	1,664,623	5,357,906	208,958	91,731	12,384	75,000	3,750	2,411,624
1890	442,368	1,819,682	1,910,666	12,029,217	541,315	124,102	19,360	75,000	4,125	2,917,834
1891	438,567	1,771,055	1,753,344	7,947,786	341,755	57,572	7,369	75,000	3,750	2,544,785
1892	314,041	1,691,579	1,471,674	7,916,672	316,667	40,424	4,689	250,000	11,500	2,118,571
1893	584,187	2,218,377	1,730,334	8,000,000	296,000	40,000	4,320	400,000	16,000	2,630,841
1894	662,649	2,228,846	1,404,173	8,000,000	264,000	40,000	3,800	200,000	7,000	2,341,622
1895	674,210	1,585,483	1,030,564	6,415,936	205,310	44,168	4,726	200,000	7,200	1,922,010
1896 [7]	792,631	1,626,828	1,106,243	6,438,672	193,160	204,519	22,088	400,000	15,600	2,129,722
1897	782,649	1,442,583	865,550	5,263,116	189,472	516,034	61,924	300,000	12,300	1,911,895
1898	605,528	1,569,012	925,717	5,843,767	222,063	317,423	39,360	300,000	13,800	1,806,468
1899	546,825	1,502,900	901,740	7,216,260	324,732	292,966	50,097	300,000	17,400	1,840,794
1900	465,447	1,358,143	842,143	4,994,263	219,748	244,092	40,519	300,000	13,200	1,580,963
1901	540,975	1,271,227	762,736	3,890,216	167,279	374,534	62,547	300,000	12,300	1,545,837
1902	930,481	1,279,050	677,897	3,282,270	134,573	473,754	57,798	317,705	15,250	1,815,999
1903	472,061	851,638	459,885	3,451,849	144,978	289,876	39,713	656,000	35,424	1,152,062
1904	636,615	873,949	506,890	3,913,976	168,301	401,180	51,351	906,705	46,242	1,409,399
1905	503,698	692,437	422,387	3,270,211	153,700	235,669	36,764	1,102,301	65,036	1,181,585
1906	529,753	652,796	443,901	3,307,001	188,499	235,375	45,427	1,733,477	165,742	1,313,322
1907	522,806	518,364	342,120	2,804,172	148,621	171,340	34,268	2,771,960	163,546	1,211,451
1908	659,116	503,551	266,882	2,015,010	84,630	264,994	34,979	836,411	9,311	1,084,918
1909	536,407	448,535	233,238	3,264,675	139,951	299,546	38,941	758,074	40,936	989,473
1910	522,524	475,174	256,594	2,434,476	107,117	595,795	75,666	1,247,389	67,359	1,029,260
1911	519,207	437,841	232,056	3,325,222	149,635	650,368	81,296	1,417,544	80,680	1,062,994
1912	445,794	373,940	229,973	3,523,733	158,568	449,401	74,151	1,734,493	119,680	1,028,166
1913	432,489	408,527	246,750	3,999,614	175,983	426,393	66,091	1,489,518	83,413	1,004,726
1914	495,275	345,387	190,999	2,435,692	94,992	367,790	48,916	1,067,314	54,433	884,615

Years	GOLD Value	SILVER Fine Ounces	SILVER Value	LEAD Pounds	LEAD Value	COPPER Pounds	COPPER Value	ZINC Pounds	ZINC Value	Total
1915	526,583	393,108	199,306	2,527,575	118,796	530,949	92,916	1,505,032	186,624	1,124,225
1916	428,931	462,141	304,089	4,295,725	296,405	621,732	152,946	2,572,575	344,725	1,527,096
1917	303,984	526,750	434,042	4,836,617	415,949	570,091	155,635	3,153,030	321,609	1,631,219
Totals	$22,069,474	56,349,314	$50,883,493	166,953,311	$7,363,312	11,278,318	$1,791,457	26,594,528	$1,915,055	$84,022,791
*1918	200,000	376,653	364,505	3,762,813	281,437	462,172	113,556	1,872,200	148,000	1,107,498

*Estimate.

Compiled from figures by C. W. Henderson, Statistician, U. S. Geological Survey, and other sources.

TOTAL MINERAL PRODUCTION BY YEARS FOR DELTA COUNTY

Years	GOLD Value	SILVER Fine Ounces	SILVER Value	LEAD Pounds	LEAD Value	COPPER Pounds	COPPER Value	ZINC Pounds	ZINC Value	Total
1894	$ 172	3	$ 2							$ 174
1895	77	1	1							78
1896	339	1	1							340
1897	289									289
1898	579	16	9							588
1899	207	10	6							213
1900	971	97	60							1,031
1901	517	10	6							523
1902	413	12	6							419
1903	248	8	4							252
1904	351	9	6							357
1905-1909	No production.									
1910	110	139	75							185
1911-1917	No production.									
Totals	$4,273	305	$176							$4,449
1918	No production.									

Compiled from figures by C. W. Henderson, Statistician, U. S. Geological Survey, and from other sources.

TOTAL MINERAL PRODUCTION BY YEARS FOR COSTILLA COUNTY

Years	GOLD Value	SILVER Fine Ounces	SILVER Value	LEAD Pounds	LEAD Value	COPPER Pounds	COPPER Value	ZINC Pounds	ZINC Value	Total
1874-1884	$ 216									$ 216
1885										
1886-1894	126									126
1895	139									139
1896										
1897	5,416	482	$ 289	50,048	$1,802	502	$ 60			7,567
1898	5,519	993	586			983	122			6,227
1899	806	126	76							882
1900	2,067	314	195			107	18			2,280
1901	971	153	92			235	39			1,102
1902	1,178	205	109							1,287
1903	992	179	97							1,089
1904	668	151	88							756
1905										
1906	426									426
1907-1909										
1910	2,318	9	5							2,323
1911	21,832	96	51							21,883
1912	470	3	2							472
1913	95	2	1							96
1914										
1915-1917	177	2	1							178
Totals	$43,416	2,715	$1,592	50,048	$1,802	1,827	$239			$47,049
1918	No production.									

Compiled from figures by C. W. Henderson, Statistician, U. S. Geological Survey, and from other sources.

TOTAL MINERAL PRODUCTION BY YEARS FOR CUSTER COUNTY

Years	GOLD Value	SILVER Fine Ounces	SILVER Value	LEAD Pounds	LEAD Value	COPPER Pounds	COPPER Value	ZINC Pounds	ZINC Value	Total
1872		6,051	$ 8,000							$ 8,000
1873		7,721	10,014							10,014
1874		17,005	21,732							22,854
1875		156,142	193,616							193,616
1876		38,672	44,860							44,860

Year	1	2	3	4	5	6	7	8	9	10
1877	$50,000	77,344	92,813							142,813
1878	100,000	77,344	88,946							188,946
1879	100,000	541,406	606,375							706,375
1880	100,000	665,156	764,929							864,929
1881	100,000	541,406	611,789							711,789
1882	200,000	232,031	264,515							464,515
1883	620,000	154,688	171,704							791,704
1884	350,000	185,625	206,044	$500,000	$18,500					574,544
1885	30,000	61,295	65,586	5,440,000	212,160					307,746
1886	21,600	61,295	60,682	4,500,000	207,000					289,282
1887	507	117,970	115,611	5,367,459	241,536					357,654
1888	120	3,463	3,255	4,821,143	212,130					215,505
1889	1,281	72,576	68,221	63,086	2,460					71,962
1890	114,212	119,684	125,668	1,708,729	76,893					316,773
1891	49,204	48,469	47,984	858,874	36,072					133,260
1892	325	9,635	8,382	4,963	199					8,906
1893	4,021	32,204	25,119	150,000	5,550					34,690
1894	148	1,137	716	150,000	4,950					5,814
1895	68	88,632	57,611	139,768	4,473	$4,099	$439			62,591
1896	42	60,122	40,883	82,105	2,463	1,109	120			43,508
1897	2,129	26,842	16,105	2,101,041	75,637	874	105			93,976
1898	723	24,319	14,348	996,877	37,881	1,475	183			53,135
1899	1,054	6,004	3,602	836,894	37,660	923	158			42,474
1900	20,835	82,605	51,215	709,349	31,211	2,301	382	$20,000	$880	104,523
1901	11,120	50,394	30,236	400,481	17,221	40,528	6,768			65,345
1902	23,708	28,189	14,940	94,662	3,881	32,945	4,019	40,500	1,944	48,492
1903	82,804	160,175	86,495	387,301	16,267	52,242	7,157			192,723
1904	53,453	87,375	50,676	126,593	5,444	15,068	1,929			111,502
1905	24,918	32,159	19,617			2,500	390			44,925
1906	16,318	79,480	54,046	115,960	6,610	2,725	526	971	59	77,559
1907	6,845	25,995	17,157	103,585	5,490	8,420	1,684			31,176
1908	7,183	13,156	6,973	120,330	5,054	243	32			19,242
1909	12,774	14,796	7,694	41,721	1,794	700	91	89,593	4,838	27,191
1910	9,889	7,767	4,144	14,796	651	3,882	493	6,796	367	15,544
1911	5,560	13,179	6,985	17,511	788	1,640	205			13,538
1912	16,898	25,426	15,637	10,444	470	2,006	331			33,336
1913	14,684	11,313	6,883	5,273	232	4,052	628			22,427
1914	3,365	15,975	8,834	9,692	378	3,481	463	4,470	228	13,268
1915	4,098	31,633	16,038	89,808	4,221	12,640	2,212	30,411	3,771	30,340
1916	6,309	36,971	24,327	123,536	8,524	44,004	10,825	10,970	1,470	51,455
1917	7,066	88,687	73,078	228,303	19,634	88,216	24,083			123,861
Totals	$2,173,211	4,239,511	$4,234,085	30,300,274	$1,303,434	331,173	$64,345	203,711	$13,557	$7,788,632
*1918	4,000	106,770	103,323	53,306	3,987	45,922	11,283			122,593

*Estimate.

Compiled from figures by C. W. Henderson, Statistician, U. S. Geological Survey, and from other sources.

TOTAL MINERAL PRODUCTION BY YEARS FOR DOLORES COUNTY

Years	GOLD Value	SILVER Fine Ounces	SILVER Value	LEAD Pounds	LEAD Value	COPPER Pounds	COPPER Value	ZINC Pounds	ZINC Value	Total
1833-78	$ 1,500	7,734	$ 8,662	10,000	$ 410	4,301	$ 800			11,372
1879	3,500	30,938	35,579	100,000	5,000	29,000	6,206			50,285
1880	5,000	69,610	78,659	200,000	9,600	44,000	8,008			101,267
1881	10,000	85,078	96,989	200,000	9,800	54,000	10,314			127,103
1882										
1883	5,000	193,360	214, 80	20, 00	8,600	100,000	16,500			244,730
1884	1,500	54,141	60,097	152,000	5,624					67,221
1885	1,00	70,000	74,900	100,000	3,900					82,800
86	8,561	75,836	75,078	792,000	36,432					71
1887	9, 47	118,262	115,897	1,000,000	45,000	34,000	4,692			175,332
1888	17,470	123,852	116,421	1,000,000	44,000					177,891
89	77,825	618,615	581,498	2,000,000	78,000					737,323
90	156,297	848,785	891,224	2,000,000	90,000					1,137,521
91	122,631	699,888	692,889	931,326	40,047					567
1892	235,669	1,285,179	1,118,106	3,083,168	123,327	13,043	1,513			1,478,615
93	442,105	2,675,238	2,086,686	4,500,000	166,500	10,000	180			2,696,371
1894	192,626	1,153,325	726,595	2,000,000	66,000	30,000	2,850			988,071
1895	52,552	399,283	259,534	313,824	10,042	64,151	6,864			328,992
1896	10,659	240,393	163,467	1,100,000	33,00,0			30,000	$ 1,1 D	208,296
97	43,469	179,901	107,941	1,093,840	348	39,654	4,758			195,546
98	88,847	273,346	273,374	686,597	26,091	149,647	18,556	400,000	18,400	4203
99	66,847	257,052	154,231	2,046,232	92,080	44,509	7,611	100,000	5,800	326,569
1900	50,125	159,318	98,777	210,380	9,257	36,009	5,9 8	220,000	9,680	173,817
1901	22,303	111,632	66,979	367,057	15,783	13,106	2,189	250,000	10,250	117,504
902	47,458	121,311	64,295	388,806	15,941	15,054	37	248,680	11,937	141,468
1903	43,262	103,096	55,672	143,417	6,024	147,588	20,220			125,178
04	53,783	108,301	62,815	181,229	7,793	25,392	3,250	18,196	928	128,569
05	34,766	76,526	46,681	840,319	9,395	119,821	18,692	555,266	32,820	754
06	9,398	34,290	23,317	118,229	6,739	199,379	3,848	883,553	53,896	330
907	11,689	33,037	21,804	54,547	2,891	99,495	19,899			56,283
98	37,238	163,563	86,688	947,462	39,814	42,495	5,609	509,184	23,932	193,281
99	22,266	103,646	53,896	462,373	19,882	43,538	5,660	167,574	9,049	110,753
1910	15,327	88,309	47,687	127,909	5,628	97,063	12,327	87,000	4,698	85,667
1911	7,565	56,202	29,787	701,244	31,556	3,288	411	525,333	29,944	99,263
1912	7,556	100,288	61,677	1,212,400	54,558	689,915	113,837	812,029	56,030	293,658

Years	GOLD Value	SILVER Fine Ounces	SILVER Value	LEAD Pounds	LEAD Value	COPPER Pounds	COPPER Value	ZINC Pounds	ZINC Value	Total
1913	12,432	178,816	108,005	3,079,341	135,491	801,819	124,282	2,596,232	145,389	525,599
1914	7,973	86,526	47,849	492,023	19,189	350,278	46,587	366,549	18,694	140,292
1915	11,932	127,933	64,862	268,447	12,617	1,032,480	180,684	35,936	4,456	274,551
1916	7,426	77,280	50,850	588,333	40,595	419,500	103,197	182,306	24,429	226,497
1917	5,213	88,222	72,695	1,772,221	152,411	519,916	141,937	1,701,353	173,538	545,794
Totals	$1,962,948	11,468,112	$8,996,793	35,465,224	$1,548,495	5,272,441	$934,828	9,690,171	$635,040	$14,078,104
*1918	2,600	40,604	40,294	620,421	46,404	667,830	164,086	619,850	49,000	302,384

*Estimate.
Compiled from figures by C. W. Henderson, Statistician, U. S. Geological Survey, and other sources.

TOTAL MINERAL PRODUCTION BY YEARS FOR CONEJOS COUNTY

Years	GOLD Value	SILVER Fine Ounces	SILVER Value	LEAD Pounds	LEAD Value	COPPER Pounds	COPPER Value	ZINC Pounds	ZINC Value	Total
1861-1884	No production.									
1885	$ 277									$ 277
1886-1893										
1894	171	1	$ 1							.172
1895										
1896	639	17	12							651
1897	1,054	98	59							1,113
1898	18,355	29,777	17,568							35,923
1899	6,263	22,987	13,792							20,055
1900	2,832	1,014	629	2,200	$ 97	4,527	$752			4,310
1901	1,178	102	61	1,200	52	210	35			1,326
1902	1,261	81	43			78	10			1,314
1903	1,220	46	25							1,245
1904	827	52	30							857
1905	2,894	900	549							3,443
1906	No production.									
1907-1917	1,474	747	509							1,983
Totals	$38,445	55,823	$33,278	3,400	$149	4,815	$797			$72,669
1918	No production.									

Compiled from figures by C. W. Henderson, Statistician, U. S. Geological Survey, and from other sources.

TOTAL MINERAL PRODUCTION BY YEARS FOR DOUGLAS COUNTY

Years	GOLD Value	SILVER Fine Ounces	SILVER Value	LEAD Pounds	LEAD Value	COPPER Pounds	COPPER Value	ZINC Pounds	ZINC Value	Total
1861-84	No production.									
	$1,420	70	$ 75							$1,495
85										....
86										81
87	475	10	6							24
										97
98	124	24	14							09
1899	83	24	15							67
1900	62	10	6							42
01	103	10	5							32
1902	62									
03	41	2	1							4
04	289	5	3							49
05										
06	4									11
07	49									
08	131									83
09										56
1910	83									75
1911	166									
1912	75									
1913	547	2	1							547
1914	140									40
1915	596	4	2							98
1916-17	No production.									
	$4,450	161	$128							$4,578
1918	No production.									

Compiled from figures by C. W. Henderson, Statistician, U. S. Geological Survey, and from other sources.

TOTAL MINERAL PRODUCTION BY YEARS FOR EAGLE COUNTY

Years	GOLD Fine Ounces	GOLD Value	SILVER Fine Ounces	SILVER Value	LEAD Pounds	LEAD Value	COPPER Pounds	COPPER Value	ZINC Pounds	ZINC Value	Total
1879	No...	$				$					$
1880	2,000		38,672	$ 44,473	800,000	40,000					86,473
1881	4,000		79,344	89,659	1,600,000	76,800					170,459
1882	5,000		98,680	112,495	2,000,000	98,000					215,495
1883	70,000		232,031	257,554	16,000,000	688,000					1,015,554

Year										
1884	30,000	154,687	171,703	6,600,000	244,200					445,903
1885	33,000	170,156	182,067	5,950,000	232,050					447,117
1886	423,517	569,637	563,941	2,000,000	92,000					1,079,458
1887	219,594	254,078	248,996	1,112,905	50,081					518,671
1888	142,002	193,489	181,880	2,370,090	104,284					428,166
1889	92,220	170,551	160,318	2,112,280	82,379					334,917
1890	68,862	75,265	79,028	1,000,000	45,000					192,890
1891	153,453	280,168	277,366	3,776,230	162,378					53 37
1892	139,299	347,954	302,720	5,89	210,371					30
1893	168,867	187,658	146,372	60	185,000					500,240
1894	55,521	62,543	39,40	2,000,000	66,000					160,923
1895	30,900	53,421	34,724	1,770,215	56,647					122,271
1896	16,442	65,824	44,760	210,717	6,322	2,044	221			67,775
1897	34,767	46,046	27,628	1,144,013	41,184	2,200	264			103,843
1898	30,571	70,783	41,762	1,851,512	70,357	71,049	8,810			151,500
1899	46,094	44,393	26,636	1,187,930	53,457	5,86	1,005			127,192
1900	103,598	234,674	145,498	3,679,828	161,912	89,64	59,603			471,491
1901	97,376	175,181	105,109	2,775,291	119,338	37,914	26,372	20,000		348,195
1902	31,956	45,336	24,028	832,846	34,147	50,14	18,316		880	108,447
1903	16,040	27,054	14,609	677,730	28,465	32,63	4,502			63,616
1904	30,075	27,348	15,862	375,207	16,134	32,409	4,148	605,612	35,731	66,219
1905	46,891	46,487	28,357	156,723	7,366	29,331	4,576	1,426,029	86,988	122,921
1906	51,561	94,912	64,540	307,755	17,542	130,233	25,135	429,198	25,323	425,766
1907	53,641	70,586	46,587	193,680	10,266	14,270	2,854			138,671
1908	58,131	86,715	45,959	11,204	71	66,144	8,731			113,292
1909	53,308	125,214	51	152,280	6,548	286,885	37,295	740,408	39,982	202,244
1910	26,231	88,313	47,689	397,409	17,486	209,551	26,613	4,147,945	223,989	341,008
1911	41,160	116,109	61,538	855,889	38,515	66,608	8,326	5,097,597	290,563	440,102
1912	49,94	163,735	100,697	1,240,156	55,807	147,176	24,284	5,659,261	390,489	620,571
1913	41,220	301,380	182,034	1,351,205	59,453	41,368	6,412	6,683,643	374,284	663,403
1914	47,194	127,080	70,275	1,177,385	45,918	28,105	3,738	7,522,098	383,627	550,752
1915	95,426	177,550	90,018	1,394,43	65,520	60,086	10,515	11,141,750	1,381,577	1,643,056
1916	96,036	222,126	146,159	1,517,362	104,698	112,610	27,702	28,438,052	3,810,699	4,185,294
1917	41,187	136,023	112,083	2,426,988	208,721	53,136	14,506	23,715,412	2,418,972	2,795,460
Totals	$2,745,464	5,461,203	$4,399,638	83,266,163	$3,602,817	2,059,043	$ 23,98	95,627,005	$9,463,104	$20,534,951
1918*	37,000	261,748	253,294	953,414	71,310	417,631	102,612	21,125,500	1,670,000	2,134,216

*Estimate.

Compiled from figures by C. W. Henderson, Statistician, U. S. Geological Survey, and from other sources.

TOTAL PRODUCTION OF EL PASO COUNTY, COLO., 1913-1914

Year	Copper Quantity, Pounds	Copper Value	Total Value
1913	10,632	$1,648	$1,648
1914	2,644	352	352
Totals	13,276	$2,000	$2,000

Compiled from figures by C. W. Henderson, Statistician, U. S. Geological Survey.

TOTAL MINERAL PRODUCTION BY YEARS FOR FREMONT COUNTY

Years	GOLD Value	SILVER Fine Ounces	SILVER Value	LEAD Pounds	LEAD Value	COPPER Pounds	COPPER Value	ZINC Pounds	ZINC Value	Total
1881		11,602	$13,110							$13,110
1882		15,469	17,635							17,635
1883		15,469	17,171							17,171
1884-1886										
1887	$ 186	474	465	5,930	$ 267					958
1888	10,841	21,683	20,382	466,538	18,195	5,317	718			50,136
1889	No production.									
1890-1893	76	323	203							279
1894	18									18
1895										
1896	915	15	10							925
1897	12,877	1,525	915							13,792
1898	8,702	1,270	749	2,101	80					9,531
1899	9,405	3,974	2,384	11,443	515	6,698	1,145			13,449
1900	8,309	2,199	1,363	8,282	365	6,725	1,116			11,153
1901	6,449	933	560	33,945	1,460	15,907	2,656			11,125
1902	7,379	515	273	2,836	116	22,300	2,721	22,825	$ 1,096	11,585
1903	6,346	223	120	2,091	88	20,777	2,846			9,400
1904	4,671	208	121	1,071	46	1,024	131			4,969
1905										
1906	77	79	54					568,508	34,679	34,810
1907	302	561	370			30,330	6,066			6,738
1908	91	4	2							93
1909	85					677	88			173
1910								18,072	976	976

TOTAL MINERAL PRODUCTION BY YEARS FOR GILPIN COUNTY

Years	GOLD Value	SILVER Fine Ounces	SILVER Value	LEAD Pounds	LEAD Value	COPPER Pounds	COPPER Value	ZINC Pounds	ZINC Value	Total
1911	178	1,345	713	19,904	896	13,976	1,747	140,526	8,010	11,544
1912	253	3,439	2,115	55,956	2,518	35,903	5,924	447,507	30,878	41,688
1913	92	78	47	4,591	202	4,677	5,725	7,161	401	1,467
1914	1,476	1,066	589	308	12	191,917	25,525			27,602
1915	674	3,168	1,606	30,894	1,452	127,303	22,278	228,170	28,293	54,303-
1916	786	4,529	2,980	31,710	2,188	101,041	24,856			30,810
1917	590	664	597			59,857	16,341			17,478
Totals --	$80,778	90,815	$84,484	677,600	$28,400	644,429	$114,883	1,432,769	$104,333	$412,878
1918	Production unknown.									

Compiled from figures by C. W. Henderson, Statistician, U. S. Geological Survey, and from other sources.

Years	GOLD Value	SILVER Fine Ounces	SILVER Value	LEAD Pounds	LEAD Value	COPPER Pounds	COPPER Value	ZINC Pounds	ZINC Value	Total
1858	No production.									
1859	$ 241,918	5,943	$ 8,082							$ 250,000
1860	870,903	21,553	29,097							900,000
1861	725,753	18,231	24,247							750,000
1862	1,161,204	28,738	38,796							1,200,000
1863	1,548,222	38,460	51,728							1,600,000
1864	1,741,806	43,267	58,194							1,800,000
1865	1,451,505	36,272	48,495							1,500,000
1866	725,753	18,108	24,247							750,000
1867	967,670	24,308	32,230							1,000,000
1868	1,640,000	93,211	123,730			50,000	11,500			1,775,230
1869	2,690,000	86,340	114,400			100,000	24,250			2,828,650
1870	2,120,000	65,910	87,528			180,000	38,124			2,245,652
1871	3,237,346	59,229	78,478			180,000	43,416			3,359,240
1872	2,083,611	52,911	69,9			200,000	71,120			2,224,679
1873	1,393,931	35,907	46,571	25,000	$ 1,500	200,000	56,000			1,498,002
1874	1,525,447	39,418	50,376	50,000	3,000	252,050	55,451			1,634,274
1875	1,355,566	62,6	77,711	50,000	2,900	193,665	43,962			1,520,139
1876	1,990,002	89,365	103,663	50,000	3,050	250,000	52,590			2,149,215
1877	2,086,871	93,714	112,457	50,000	2,750	300,000	57,000			2,259,078
1878	2,155,708	96,806	111,327	50,000	1,800	300,000	49,800			2,318,635
1879	2,260,000	232,031	259,875	100,000	4,100	300,000	55,800			2,579,775
1880	2,380,000	232,031	266,836	100,000	5,000	300,000	64,200			2,716,036
1881	1,850,000	201,094	227,236	100,000	4,800	300,000	54,600			2,136,636
1882	1,600,000	201,094	229,247	100,000	4,900	400,000	76,400			1,910,547

TOTAL MINERAL PRODUCTION BY YEARS FOR GILPIN COUNTY—Continued

Years	GOLD Value	SILVER Fine Ounces	SILVER Value	LEAD Pounds	LEAD Value	COPPER Pounds	COPPER Value	ZINC Pounds	ZINC Value	Total
1883	$ 1,650,000	154,688	$ 171,704	100,000	$ 4,300	200,000	$ 33,000	......	......	$ 1,859,004
1884	1,950,000	278,438	309,066	128,411	4,751	600,000	78,000	......	......	2,341,817
1885	2,051,000	300,000	321,000	128,411	5,008	300,000	32,400	......	......	2,409,408
1886	1,337,061	101,784	100,766	200,000	9,200	300,000	33,300	......	......	1,480,327
1887	1,134,476	266,281	260,955	228,622	10,288	600,000	82,800	......	......	1,488,519
1888	1,250,756	174,364	163,902	1,288,825	56,708	400,000	67,200	......	......	1,538,566
1889	1,054,065	313,071	294,287	1,441,926	55,065	250,110	33,765	......	......	1,437,182
1890	805,236	292,495	307,120	1,130,453	50,870	620,927	96,865	......	......	1,260,091
1891	938,016	232,001	229,681	779,837	33,533	558,298	71,462	......	......	1,272,692
1892	1,358,157	134,462	116,982	2,233,158	89,286	538,988	62,523	......	......	1,626,948
1893	1,218,626	135,850	105,963	2,000,000	74,000	400,000	64,800	......	......	1,463,389
1894	1,915,863	228,927	144,224	2,200,000	72,600	400,000	38,000	......	......	2,170,687
1895	1,196,319	190,256	123,666	844,037	27,009	209,414	22,407	......	......	1,369,401
1896	1,534,358	295,182	200,724	1,943,756	58,462	435,838	47,071	......	......	1,840,615
1897	2,086,471	374,417	224,650	2,007,698	72,277	1,018,595	122,231	......	......	2,505,629
1898	1,983,514	305,687	180,355	1,216,338	46,221	633,707	78,580	......	......	2,288,670
1899	1,996,061	340,653	204,391	1,312,312	59,054	1,037,421	177,399	......	......	2,436,905
1900	1,655,502	236,400	146,568	735,773	32,374	799,478	132,713	......	......	1,967,157
1901	1,638,966	271,638	162,983	670,018	28,811	731,194	122,109	......	......	1,952,869
1902	1,551,035	303,638	160,928	497,366	20,392	765,516	193,393	......	......	1,925,748
1903	1,346,113	375,238	202,629	945,975	39,731	611,988	83,842	......	......	1,672,315
1904	1,403,85	318,406	184,675	859,293	36,950	638,945	81,785	......	......	1,707,275
1905	1,450,63	340,901	207,950	519,841	24,433	512,276	79,915	33,090	$ 1,952	1,764,283
1906	1,115,92	242,478	64,885	510,791	29,115	638,002	123,134	46,000	2,806	1,435,842
1907	98,88	209,347	138,169	611,060	32,386	874,060	174,812	......	......	1,283,855
1908	1,075,808	187,030	99,126	538,143	22,602	636,671	84,001	......	......	1,281,537
1909	887,311	172,010	89,445	664,581	28,577	499,146	64,889	......	......	1,070,222
1910	687,902	132,635	71,623	575,477	25,321	534,244	67,849	......	......	852,695
1911	778,774	292,659	155,109	1,239,356	55,771	950,240	118,780	23,088	1,316	1,109,750
1912	904,505	316,205	194,466	1,351,600	60,822	1,025,770	169,252	25,377	1,751	1,330,96
1913	687,101	273,207	165,017	1,210,341	53,255	837,974	129,886	8,589	481	1,035,740
1914	573,553	145,237	80,316	499,718	19,489	726,579	96,635	12,980	662	770,655
1915	562,878	125,665	63,712	591,127	27,783	476,383	83,367	11,000	1,364	739,104

TOTAL MINERAL PRODUCTION BY YEARS FOR GUNNISON COUNTY

Years	GOLD	SILVER		LEAD		COPPER		ZINC		Total
	Value	Fine Ounces	Value	Pounds	Value	Pounds	Value	Pounds	Value	
1861-1872	No production.									
1873	$ 5,000									5,000
1874-1878	No production.									
1879		19,336	$ 21,656							21,656
1880		222,031	255,336	100,000	$ 5,000					260,336
1881	10,000	309,375	349,594	360,000	17,280					376,874
1882	100,000	386,719	440,860	360,000	17,640					558,500
1883	100,000	502,734	558,035	500,000	21,500	30,000	$ 4,950			684,485
1884	60,000	386,719	429,258	2,000,000	74,000					563,258
1885	40,000	144,323	154,426		92,820					287,246
1886	18,226	144,323	142,880	500,000	23,000					184,106
1887	50,486	172,616	169,184	451,351	20,311					239,981
1888	18,642	60,166	56,556	1,011,792	44,519	1,556				119,717
1889	39,710	48,106	45,220	485,355	18,929		75			103,934
1890	28,784	354,393	372,113	6,945,972	312,569	105,954	16,529			729,995
1891	7,402	489,268	484,375	10,340,332	444,634					936,411
1892	6,004	146,891	127,795	525,574	21,023					154,822
1893	7,728	144,577	112,770	500,000	18,500					138,998
1894	8,062	104,938	66,111	400,000	13,200					87,373
1895	36,734	114,218	74,242	201,898	6,461					117,437
1896	26,757	93,237	63,426	164,370	4,931	8,515	920			96,034
1897	40,761	103,941	62,365	1,013,114	36,472	2,770	332			139,930
1898	81,006	152,800	90,152	1,996,560	75,869	119,072	14,765			261,792
1899	70,112	132,983	79,790	1,399,336	62,970	46,186	7,898			220,770
1900	83,858	146,746	90,982	1,583,320	69,666	42,790	7,103	100,000	$ 4,400	256,009
1916	453,259	126,553	83,272	521,334	35,972	557,317	137,100	141,490		709,603
1917	397,087	112,585	92,770	815,906	70,168	544,648	148,689		14,432	723,146
Totals	$83,411,327	10,177,098	$8,197,678	33,190,484	$1,406,384	24,569,134	$3,988,077	301,614	$24,764	$97,028,230
*1918	300,000	66,986	64,824	550,916	33,726	260,036	63,891			462,441

*Estimate.

Compiled from figures by C. W. Henderson, Statistician, U. S. Geological Survey, and from other sources.

TOTAL MINERAL PRODUCTION BY YEARS FOR GUNNISON COUNTY—Continued

Years	GOLD Value	SILVER Fine Ounces	SILVER Value	LEAD Pounds	LEAD Value	COPPER Pounds	COPPER Value	ZINC Pounds	ZINC Value	Total
1901	$ 83,445	93, 33	$ 55,946	656,631	$ 38,235	53,396	$ 8,917	100,000	$ 4,100	$ 180,643
1902	103,636	22, 33	65,263	728,935	29,886	28,686	3,500	131,975	6,335	208,520
1903	48,533	65, 47	35,344	127,661	5,362	15,000	2,055	55,600	3,002	94,293
1904	26,024	115, 53	66,789	200,462	8,620	16,233	2,078	20,010	1,021	104,532
1905	28,156	53, 69	32,726	219,809	10,331	50,500	7,878	17,905	1,056	80,147
1906	8,505	70,798	48,143	248,735	14,178			158,198	9,650	159,476
1907	61,569	27,277	18,003	120,226	6,372	13,690	2,738	38,224	2,255	90,937
1908	100,032	28,664	15,192	327,612	13,760	5,481	724	147,000	6,909	136,617
1909	108,493	37,423	19,460	493,070	21,202	51,815	6,736	212,093	11,453	167,344
1910	233,972	49,189	26,562	581,841	25,601	21,024	2,670	176,815	9,548	298,353
1911	145,039	32,541	17,247	631,933	28,437	9,928	1,241	557,456	31,775	223,739
1912	125,327	29,035	17,857	306,867	13,809	8,097	1,336	483,884	33,388	191,717
1913	10,189	87,488	52,843	196,728	8,656	21,864	3,389	292,875	16,401	91,478
1914	13,033	59,036	32,647	317,974	12,401	11,188	1,488	525,000	26,775	86,344
1915	60,197	24,892	12,620	190,000	8,930	9,091	1,591	1,750,944	217,117	300,455
1916	31,553	29,023	19,097	313,217	21,612	84,679	20,831	1,964,873	263,293	356,386
1917	6,635	40,272	33,184	751,000	64,586	180,121	49,173	3,054,990	311,609	465,187
Totals	$2,112,520	5,346,744	$4,816,026	39,631,677	$1,723,272	936,636	$168,917	9,787,842	$960,087	$9,780,822
*1918	12,000	7,364	7,090	17,635	1,319	63,655	15,640	2,061,950	163,000	199,049

*Estimate.
Compiled from figures by C. W. Henderson, Statistician, U. S. Geological Survey, and from other sources.

TOTAL MINERAL PRODUCTION BY YEARS FOR HINSDALE COUNTY

Years	GOLD Value	SILVER Fine Ounces	SILVER Value	LEAD Pounds	LEAD Value	COPPER Pounds	COPPER Value	ZINC Pounds	ZINC Value	Total
1871-1874	No production.									
1875	$ 12,000	47,953	$ 59,462							71,462
1876	20,000	154,688	179,438	50,000	$ 3,050					202,488
1877	25,000	92,814	111,377	100,000	5,500					141,877
1878	20,000	154,688	117,891	200,000	7,200					205,091
1879	6,000	193,359	216,562	500,000	20,500	30,000	$ 6,420			243,062
1880	6,000	116,016	133,418	1,000,000	57,600	40,000	7,280			195,838
1881	10,000	123,750	139,838	1,200,000	57,600	40,000	7,640			214,718
1882	20,000	61,875	70,538	600,000	29,400	22,652	3,738			127,578
1883	20,000	193,359	214,628	1,000,000	43,000					281,366

Year										
1884	2,500	154,687	171,703	1,000,00	37,000	350,000	45,500			256,703
1885	2,000	16,320	17,462	00,00	3,900	46,460	5,018			28,380
1886	2,060	16,320	16,157	00,00	4,638	46,460	5,157			27,974
1887	4,308	90,355	88,548	57,63	24,638	12,027	1,660			119,154
1888	66,7	86,248	81,073	1,205,93	53,063	2,000	36			137,139
1889	1,794	6,665	15,665	20,812	9,392	17,359	2,343			29,194
1890	3,697	57,387	60,256	60,08	29,732	60,584	9,451			103,136
1891	19,869	186,850	184,982	8,308,08	357,246	8,248	1,056			563,153
1892	22,514	411,758	358,229	4,753,83	190,151	29,914	3,470			574,364
1893	88,50	385,653	300,809	3,808,111	140,900	10,000	1,080			531,539
1894	85,96	395,899	249,416	3,322,170	109,632	10,000	90			445,194
1895	243,195	483,565	314,317	5,251,014	8,62	10,000	1,070			726,614
1896	212,794	510,883	347,400	5,468,856	164,066	3,202	1,426			725,686
1897	168,171	243,437	146,062	5,550,058	199,802	685	90			515,005
1898	51,282	186,456	110,009	9,828,482	373,482	104,038	12,901			547,674
1899	38,343	155,902	93,541	10,572,353	35,56	49,676	8,495	100,000	$4,400	616,135
1900	56,470	155,485	96,401	9,377,062	412,51	29,180	4,844	126,591	5,190	574,706
1901	76,148	152,122	91,273	7,568,675	26,313	12,552	613	319,000	15,312	501,017
1902	98,348	117,177	62,104	6,213,762	84,94	8,314	1,014	106,000	5,724	431,542
1903	16,515	33,139	17,896	459,462	19,97	11,263	542			60,974
1904	10,521	46,585	27,019	1,041,222	44,773	13,187	1,688	59,089	3,014	87,015
1905	11,991	54,419	33,196	767,681	36,081	84,485	13,180	235,178	13,876	108,324
1906	24,510	87,940	59,799	753,950	42,975	6,681	12,279	238,387	2,342	141,905
1907	7,520	50,109	33,072	1,204,628	63,845	99,410	19,882			124,319
1908	2,454	29,498	15,634	280,465	11,780	188,698	24,908			54,776
1909	7,587	75,731	39,380	106,327	4,572	714,69	92,894			144,433
1910	6,320	54,422	29,388	296,182	13,032	465,472	59,115	36,439	2,077	107,855
1911	3,830	7,753	4,109	118,645	5,339	21,696	2,712	11,926	823	18,067
1912	6,811	34,722	21,354	1,257,800	56,601	53,739	8,867	54,732	3,065	94,456
1913	5,280	30,477	18,407	782,318	34,422	76,304	11,827			73,002
1914	170	5,987	3,311	5,723	23	17,098	2,274			5,978
1915	737	9,621	4,878	266,128	12,508	9,114	1,595			19,718
1916	1,346	10,030	6,600	75,638	5,219	16,248	3,997	12,575	1,685	18,847
1917	1,136	7,721	362	209,616	18,027	6,099	1,665	4,117	420	27,610
Totals	$1,425,834	5,499,825	$4,428,964	96,173,156	$3,920,004	2,801,734	$392,338	1,104,034	$57,928	$10,225,068
*1918	7,000	23,615	22,853	901,768	67,448	20,272	4,981			102,282

*Estimate.
Compiled from figures by C. W. Henderson, Statistician, U. S. Geological Survey, and from other sources.

TOTAL MINERAL PRODUCTION BY YEARS FOR GARFIELD COUNTY

Years	GOLD Value	SILVER Fine Ounces	SILVER Value	LEAD Pounds	LEAD Value	COPPER Pounds	COPPER Value	ZINC Pounds	ZINC Value	Total
1878-1884	No production.									
1885	$ 113	45	$ 48							$ 161
1886-1893	No production.									
1894	63									63
1895	153	1	1							154
1896										
1897	310	42	25							335
1898	723	17	10							733
1899	517	13	8							525
1900										
1901	351	13	8							359
1902	165	5	3							168
1903	103	3	2							105
1904	517	14	8							525
1905										
1906	55	3	2							57
1907-1909	No production.									
1910	3,603	113	61			425	$ 54			3,718
1911-1912	No production.									
1913	890	35	21			200	31			942
1914	2,403	80	44			128	17			2,464
1915	5,309	112	57			291	51			5,417
1916										
1917	721	17	14							735
Totals	$15,996	513	$312			1,044	$153			61
1918	No production.									

Compiled from figures by C. W. Henderson, Statistician, U. S. Geological Survey, and from other sources.

TOTAL MINERAL PRODUCTION BY YEARS FOR GRAND COUNTY

Years	GOLD Value	SILVER Fine Ounces	SILVER Value	LEAD Pounds	LEAD Value	COPPER Pounds	COPPER Value	ZINC Pounds	ZINC Value	Total
1859-1895	Production credited to other counties.									
1896	$ 200									$ 200
1897	1,943	85	$ 51							1,994
1898	806	11	6							812
1899	124	13	8							132

Years	GOLD Value	SILVER Fine Ounces	SILVER Value	LEAD Pounds	LEAD Value	COPPER Pounds	COPPER Value	ZINC Pounds	ZINC Value	Total
1900	3,762	21	13							3,775
1901	1,034	30	18							1,052
1902	1,302	24	13							1,315
1903	1,426	12	6							1,432
1904	641	13	8			1,114	143			792
1905	31	22	13			1,680	262			306
1906	No production.									
1907	18									18
1908	556	72	38	690	$29	1,561	206			829
1909	1,183	9	5							1,188
1910	No production.									
1911-1913	No production.									
1914						56	7			7
1915	3	1,747	966	1,563	61					1,030
1916	153	2	1							154
19[17]		134	88			760	187			275
Totals	$13,182	2,195	$1,234	2,253	$90	5,171	$805			$15,311
1918	No production.									

Compiled from figures by C. W. Henderson, Statistician, U. S. Geological Survey, and from other sources.

TOTAL MINERAL PRODUCTION BY YEARS FOR JEFFERSON COUNTY

Years	GOLD Value	SILVER Fine Ounces	SILVER Value	LEAD Pounds	LEAD Value	COPPER Pounds	COPPER Value	ZINC Pounds	ZINC Value	Total
1858-1884	No record of production until 1885.									
1885	$ 697	.5	$ 5							$ 702
1886	2,804	43	42							2,846
1887	942	5	5							947
1888-1893	No production.									
1894	2,197	10	6							2,203
1895	2,592	15	10							2,602
1896	18,486	4,590	3,121							21,607
1897	8,247	1,614	968	10,093	$363	1,602	192			9,770
1898	1,810	102	60							1,900
1899	1,364	371	211	770	35	254	43			1,653
1900	703	51	32							735
1901	310	20	12							322
1902	517	3	2			2,978	363			882
1903	248	5	3			218	30			281
1904	3,245	37	21			538	69			3,335

TOTAL MINERAL PRODUCTION BY YEARS FOR JEFFERSON COUNTY—Continued

Years	GOLD Value	SILVER Fine Ounces	SILVER Value	LEAD Pounds	LEAD Value	COPPER Pounds	COPPER Value	ZINC Pounds	ZINC Value	Total
1905	$18,088	125	$ 76			9,000	$1,404			$19,568
1906						3,150	608			608
1907		73	48			1,955	391			439
1908										
1909	16									16
1910-1917	No production.									
Totals	$62,296	7,049	$4,622	10,863	$398	19,695	$3,100			$70,416
1918	No production.									

TOTAL MINERAL PRODUCTION BY YEARS FOR LARIMER AND JACKSON COUNTIES

Years	GOLD Value	SILVER Fine Ounces	SILVER Value	LEAD Pounds	LEAD Value	COPPER Pounds	COPPER Value	ZINC Pounds	ZINC Value	Total
1858-1894	No production.									
1895	$ 320	1	$ 1							$ 321
1896	13									15
1897	2,976	3	2							3,034
1898	11,162	97	58			24,484	$ 3,036			14,233
1899	2,067	60	35			2,474	423			2,571
1900	1,633	135	81			13,806	2,292			4,003
1901	930	126	78			18,140	3,029			4,003
1902	806	73	44			24,888	3,036			3,868
1903	1,633	49	26			56,700	7,768			9,406
1904	1,178	10	5			23,028	2,948			4,132
1905		11	6			41,331	7,977			9,653
1906	904	1,136	772							
190 -1908	No production.									
1909								30,722	$1,659	1,659
1910-1915	No production.									
1916	95	199	131			6,752	1,661			1,887
1917	587	602	496			23,725	6,477			7,560
Totals	$24,304	2,502	$1,735			235,328	$38,647	30,722	$1,659	$66,345
1918	No production.									

Compiled from figures by C. W. Henderson, Statistician, U. S. Geological Survey, and from other sources.

TOTAL MINERAL PRODUCTION BY YEARS FOR LAKE COUNTY

Years	GOLD Value	SILVER Fine Ounces	SILVER Value	LEAD Pounds	LEAD Value	COPPER Pounds	COPPER Value	ZINC Pounds	ZINC Value	Total
1860-1867	$5,272,000	37,600	$50,422							$5,322,422
1868	60,000	452	600							60,600
1869	90,000	679	900							90,900
1870	65,000	465	618							65,618
1871	100,000	1,158	1,534							101,534
1872	133,000	1,540	2,036							135,036
1873	225,000	2,937	3,809							228,809
1874	213,503	2,797	3,575							217,078
1875	43,099	16,668	20,668							63,767
1876	60,000	23,203	26,915	15,000	$915					87,830
1877	55,000	458,000	549,600	1,200,000	66,000					670,600
1878	60,000	1,800,000	2,070,000	10,000,000	350,000					2,490,000
1879	90,000	8,411,132	9,420,468	43,288,000	1,774,808					11,285,276
1880	104,014	9,977,344	11,473,946	66,658,000	3,332,900					14,910,860
1881	300,000	7,966,406	9,002,039	58,464,000	2,806,272					12,108,311
1882	320,000	8,894,531	10,139,765	97,890,000	4,796,610					15,256,375
1883	400,000	9,049,219	10,044,633	111,575,000	4,797,725					15,242,358
1884	500,000	7,270,313	8,070,047	93,628,000	3,464,236	100,000	13,000			12,047,283
1885	570,000	6,441,693	6,892,612	55,522,000	2,165,358	100,000	10,800	50,000	2,150	9,640,920
1886	433,691	6,486,047	6,421,187	84,400,000	3,882,400	100,000	11,100	50,000	2,200	10,750,578
1887	243,694	5,994,324	5,874,438	92,359,103	4,156,160	200,000	27,600	50,000	2,300	10,304,192
1888	310,891	5,466,064	5,156,064	73,785,149	3,228,639	200,000	33,600	150,000	7,500	8,737,380
1889	189,397	6,150,839	5,781,781	83,785,918	3,267,651	266,489	35,967	150,000	7,500	9,282,313
1890	295,063	5,313,930	5,579,627	43,623,477	1,963,056	1,766,035	275,501	150,000	8,250	8,121,497
1891	348,419	4,793,015	4,745,085	53,444,973	2,298,134	4,544,202	581,658	150,000	7,500	7,980,796
1892	251,296	5,898,020	5,131,277	44,009,114	1,760,365	5,928,863	687,748	562,500	25,875	7,856,561
1893	902,244	6,795,454	5,300,455	35,274,889	1,342,171	5,000,000	540,000	735,000	29,400	8,114,270
1894	1,499,314	7,695,108	4,847,918	44,733,000	1,476,189	4,000,000	380,000	1,000,000	35,000	8,238,421
1895	1,386,359	9,435,413	6,133,018	38,922,572	1,245,522	2,803,550	299,980	1,265,000	45,540	9,110,419
1896	1,453,458	6,623,764	4,504,160	31,993,777	959,813	4,071,761	439,750	642,000	25,038	7,382,219
1897	2,063,858	5,451,317	3,270,790	23,700,908	853,233	3,146,802	377,616	2,201,500	90,262	6,655,759
1898	2,073,036	7,068,727	4,170,550	35,945,006	1,365,910	5,543,954	687,450	2,673,500	122,981	8,419,927
1899	2,196,498	7,230,118	4,338,071	48,598,720	2,186,942	3,202,828	547,684	10,575,240	613,364	9,882,559
1900	2,529,512	6,967,279	4,319,713	62,599,654	2,754,385	2,728,553	452,940	14,441,000	635,404	10,691,954
1901	1,776,132	6,830,084	4,098,050	56,359,708	2,423,467	1,930,556	322,403	23,167,140	949,853	9,569,905

TOTAL MINERAL PRODUCTION BY YEARS FOR LAKE COUNTY—Continued

Years	GOLD Value	SILVER Fine Ounces	SILVER Value	LEAD Pounds	LEAD Value	COPPER Pounds	COPPER Value	ZINC Pounds	ZINC Value	Total
1902	$ 1,203, 94	5,641,857	$ 2,990,184	39,450,178	$ 1,617,457	2,617,457	$ 318,562	47,637,490	$ 2,286,600	$ 8,416,727
1903	1,339, 94	4,973,033	2,685,438	36,353,239	1,526,836	2,556,583	350,252	75,566,000	4,134,564	10,037,064
1904	1,186, 81	5,085,151	2,949,388	47,180,865	2,028,777	3,734,593	478,028	58,254,353	2,970,972	9,614,016
1905	1,180, 61	4,033,762	2,460,595	51,162,040	2,404,616	4,486,115	699,834	70,238,634	4,144,079	10,889,525
1906	1,508,410	3,890,338	2,645,430	47,456,964	2,705,047	2,092,735	403,898	70,198,462	4,282,106	11,544,891
1907	1,064,690	4,154,913	2,742,243	32,519,796	1,723,549	2,679,510	535,902	67,247,381	3,967,595	10,033,979
1908	1,228,449	2,893,496	1,533,553	19,646,007	825,132	4,674,502	617,034	23,188,080	1,089,840	5,294,008
1909	1,435,431	3,423,642	1,780,294	21,073,992	906,182	5,182,600	733,739	38,637,315	2,086,415	6,882,061
1910	1,213,134	3,322,015	1,793,888	19,249,503	846,978	3,645,157	462,935	56,367,445	3,043,842	7,360,777
1911	1,133,422	3,007,296	1,593,867	18,499,089	832,459	4,017,504	502,188	71,610,456	4,081,796	8,143,752
1912	1,103,230	3,000,397	1,845,244	26,234,244	1,180,541	2,065,800	340,857	105,945,783	7,310,259	11,780,131
1913	1,023,631	3,400,318	2,053,792	29,286,183	1,288,592	1,923,987	298,218	93,872,857	5,255,200	9,919,433
1914	1,571,451	3,810,830	2,107,389	26,784,615	1,044,600	2,382,910	316,927	78,763,334	4,016,930	9,057,297
1915	2,246,152	2,571,002	1,303,498	20,957,404	984,998	1,803,423	315,599	72,491,178	8,989,154	13,839,401
1916	1,, 80	2,931,281	1,928,783	21,719,392	1,498,668	2,621,675	644,932	76,785,567	10,289,266	16,082,059
1917	1,175,219	2,184,000	1,799,616	18,301,802	1,573,955	2,182,623	595,856	60,254,333	6,145,942	11,290,558
Totals	$47,948,307	222,898,971	$181,660,417	1,868,244,281	$81,717,218	94,294,485	$13,279,567	1,126,043,548	$76,704,527	$401,310,036
*1918	88,000	2,353,530	2,277,560	24,166,600	1,800,000	1,628,000	400,000	39,985,000	3,090,000	8,447,560

*Estimate.
Compiled from figures by C. W. Henderson, Statistician, U. S. Geological Survey, and from other sources.

TOTAL MINERAL PRODUCTION BY YEARS FOR LA PLATA AND MONTEZUMA COUNTIES

Years	GOLD Value	SILVER Fine Ounces	SILVER Value	LEAD Pounds	LEAD Value	COPPER Pounds	COPPER Value	ZINC Pounds	ZINC Value	Total
1878	$ 1,000	1,934	$ 2,224	......	......	......	......	......	......	$ 3,224
1879	2,500	3,867	4,331	......	......	......	......	......	......	6,831
1880	5,000	7,734	8,894	......	......	......	......	......	......	13,894
1881	5,000	7,734	8,739	......	......	......	......	......	......	13,739
1882	10,000	23,203	26,451	......	......	......	......	......	......	36,451

Year								Total
1883	13,000	3,867	4,292					17,292
1884	500	4,641	5,152					5,652
1885	5,000	5,000	350					10,350
1886	10,225	4,671	983	100,000	$4,600			19,449
1887	12,473	7,126		42,210	1,899			21,355
1888	3,574	2,294	2,156					5,730
1889	45	118	1,051					5,516
1890	329	2,011	2,112					5,841
1891	23,054	3,207	3,175					26,229
1892	34,881	385	2,901					37,782
1893	37,872	4,928	3,844					41,716
1894	114,264	417,465	263,003					377,267
1895	3,682	99	64					3,746
1896	10,741	41	28					0,697
1897	36,944	1,514	68	857	31	420	50	37,933
1898	63	5,219	3,079	8,407	319	2,568	318	42,369
1899	41,092	3,389	2,033	3,176	143	211	36	43,304
1900	24,927	7,187	4,456	14,500	638	350	58	30,079
1901	30,819	5,538	3,353	6,197	266	132	22	34,460
1902	127,182	7,416	3,930	2,156	88	3,143	383	131,583
1903	145,331	7,716	4,167	3,017	127	810	111	149,736
1904	130,200	31,086	18,030	2,177	94	1,473	189	148,513
1905	254,007	93,258	56,887	2,610	29	2,293	456	311,379
1906	304,633	121,721	82,770	2,228	127	445	86	387,616
9017	413,034	217,579	143,602	340	18	708	142	556,796
1908	101,584	597	37,944	748	31	458	60	139,619
1909	127,205	74,160	38,563	2,980	128	484	63	165,959
1910	399,608	141,752	76,546	273	12	362	46	476,212
1911	286,953	69,444	36,805	1,511	68	73,911	9,239	333,065
1912	135,391	47,948	29,488	6,756	304	918	151	165,334
1913	312,891	121,122	73,158	4,455	196	113	17,654	403,899
1914	126,498	60,244	35	11,410	445	68	3,463	163,721
1915	72,024	46,472	23,561	23,532	1,106	4,114	720	97,411
1916	33,055	29,370	19,332	6,667	460	15,142	3,257	56,572
1917	27,952	15,512	12,782	3,745	322	28,348	35	48,791
Totals	$3,470,943	1,683,564	$1,060,083	247,952	$11,451	276,855	$44,707	$4,587,184
1918*	10,000	689	6,473	3,559	262	777	186	16,921

*Estimate.
Compiled from figures by C. W. Henderson, Statistician, U. S. Geical Survey, and from other sources.

TOTAL MINERAL PRODUCTION BY YEARS FOR HUERFANO COUNTY

Years	GOLD Value	SILVER Fine Ounces	SILVER Value	LEAD Pounds	LEAD Value	COPPER Pounds	COPPER Value	ZINC Pounds	ZINC Value	Total
18 5-1885	No production.									
1886										$ 116
1887-1893	No production.									
1894	304	1	$ 1							305
1895	87									87
1896	109									109
1897	723	67	100	1,067	$38	92	$11			872
1898	145	40	24							169
1899	124	5	3							127
1900	124	2	12							136
1901	83	10	6							89
1902	847	260	138							985
1903										
1904										
1905	269	617	376							645
1906	475	56	38							513
1907	68									68
1908-1917	No production.									
Totals	$3,474	1,176	$698	1,067	$38	92	$11			$4,221
1918	No production.									

Compiled from figures by C. W. Henderson, Statistician, U. S. Geological Survey, and from other sources.

TOTAL MINERAL PRODUCTION BY YEARS FOR LAS ANIMAS COUNTY

Years	GOLD Value	SILVER Fine Ounces	SILVER Value	LEAD Pounds	LEAD Value	COPPER Pounds	COPPER Value	ZINC Pounds	ZINC Value	Total
1866-1886	No production.									
1887	$1,122	8	$ 8							$1,130
1888-1896	No production.									
1897	641	9	5							646
1898	124	3	2							129
1899	207									209
1900-1917	No production.									
Totals	$2,094	20	$15							$2,109
1918	No production.									

Compiled from figures by C. W. Henderson, Statistician, U. S. Geological Survey, and from other sources.

TOTAL MINERAL PRODUCTION BY YEARS FOR MINERAL COUNTY

Years	GOLD Value	SILVER Fine Ounces	SILVER Value	LEAD Pounds	LEAD Value	COPPER Pounds	COPPER Value	ZINC Pounds	ZINC Value	Total
1889										
1890										
1891	$ 10,055	378,899	$ 374,382	354,854	$ 15,259					$ 399,696
1892	87,219	2,391,514	2,080,617	3,000,000	120,000					2,287,836
1893	53,252	4,897,684	3,820,194	7,500,000	277,500					4,150,946
1894	40,336	1,866,927	1,176,164	6,500,000	214 50					1,431,000
1895	114,482	1,423,038	924,975	8,220,876	38					1,302,525
1896	52,238	1,560,865	1,061,388	6,021,109	180,633					1,294,259
1897	61,338	3,070,576	1,842,346	6,090,673	218,904	1,500				2,122,758
1898	46,383	4,177,184	2,464,539	5,453,104	209,218	14,729	$ 1,826	200,000	$ 9,200	2,729,166
1899	91,671	3,796,899	2,278,139	5,677,162	255,472	20,223	3,458	100,000	5,800	2,634,540
1900	209,387	2,280,038	1,413,623	14,951,956	657,886	2,614	84	450,000	19,800	2,301,130
1901	102,813	1,816,023	1,089,614	10,519,895	452,355	1,007	68	1,800,000	73,800	1,718,750
1902	112,838	1,923,973	1,019,706	9,291,358	380,946			2,047,555	98,283	1,611,773
1903	178,961	1,608,788	868,746	8,600,646	361,227	13	18	2,634,000	142,236	1,551,188
1904	222,864	1,664,633	965,487	13,346,436	573,897	1,337	11	4,402,697	224,538	1,986,957
1905	21,584	1,193,442	2R .00	11,880,797	558,397	07	17	2,515,628	148,422	1,651,830
1906	176,150	1,254,058	82 29	14,086,356	848,522			2,892,061	176,416	2,053,847
1907	142,803	1,246,961	82 94	12,980,288	687,955	12,711	82	2,691,216	158,781	1,815,076
1908	127,549	830,951	0?4	8,238,025	345,997	41	5	1,100,107	51,705	965,660
1909	108,825	891,185	463,416	9,036,816	388,583	17,401	2,262	1,897,296	98,134	1,061,220
1910	121,181	773,722	417,810	8,246,000	362,824	29,031	3,687	2,421,926	130,784	1,036,286
1911	179,196	545,319	289,019	7,674,556	345,355	33,384	4,173	1,258,561	71,738	889,481
1912	86,002	714,909	439,669	5,730,222	257,860	23,885	3,941	308,681	21,299	808,771
1913	50,282	805,343	486,427	3,398,264	149,528	31,647	4,905	454,875	25,473	716,615
1914	19,403	615,734	340,501	1,401,795	54,670	32,586	4,334			418,809
1915	33,039	291,807	147,946	2,382,128	111,960	8,943	1,565	85,984	10,662	305,172
1916	31,124	373,956	246,063	1,295,087	158,361	13,138	3,232	240,575	32,237	471,017
1917	10,101	361,517	297,890	1,305,744	112,294	19,297	5,268	54,971	5,607	431,160
Totals	$2,686,377	42,755,945	$27,352,818	194,974,241	$8,561,171	263,714	$42,186	27,426,128	$1,504,916	$10,747,468
1918*	12,000	652,083	631,034	962,211	71,968	3,877	92			715,954

*Estimate.
Compiled from figures by C. W. Henderson, Statistician, U. S. Geological Survey, and from other sources.

TOTAL MINERAL PRODUCTION BY YEARS FOR MONTROSE COUNTY

Years	GOLD Value	SILVER Fine Ounces	SILVER Value	LEAD Pounds	LEAD Value	COPPER Pounds	COPPER Value	ZINC Pounds	ZINC Value	Total
1886	$ 281	3	$ 3							$ 284
1887	500	9	9							509
1888	12,000									12,000
1889-1893	No production.									
1894	2,202	16	10							2,212
1895	1,181	11	7							1,188
1896	1,945	17	12							1,957
1897	6,552	851	511							7,063
1898	2,708	6,290	3,711			34,664	$ 4,298			10,717
1899	723	46,119	27,671			75,006	12,826			41,220
1900	1,633	19,652	12,184			32,026	5,316			19,133
1901	1,550	101,359	60,815			55,944	9,343			71,708
1902	5,953	3,149	1,669	64	$3	2,505	306			7,931
1903	2,811	2,061	1,113			10,920	1,496			5,420
1904	1,488	1,067	619			7,476	957			3,064
1905	114	3	2							116
1906	314	9	6							320
1907	No production.									
1908-1911	No production.									
1912	687	10	6							693
1913	940	434	262			24,058	3,729			4,931
1914	446	517	286			32,414	4,311			5,043
1915	1,277	1,073	544			57,330	10,031			11,852
1916	10	1,132	745			100,008	24,602			25,337
1917	944	666	549			21,275	5,808			7,301
Totals	$46,259	184,448	$110,734	64	$3	453,616	$83,023			$240,019
1918	No production.									

Compiled from figures by C. W. Henderson, Statistician, U. S. Geological Survey, and from other sources.

TOTAL MINERAL PRODUCTION BY YEARS FOR OURAY COUNTY

Years	GOLD Value	SILVER Fine Ounces	SILVER Value	LEAD Pounds	LEAD Value	COPPER Pounds	COPPER Value	ZINC Pounds	ZINC Total	Total
1874-1877	No production.									
1878	$ 5,000	38,672	$ 44,473							49,473
1879	8,500	38,672	43,313	198,781	$ 8,150					59,963
1880	8,500	69,610	80,052	200,000	10,000					98,552
1881	55,000	85,078	96,138	230,000	11,040	100,000	$ 18,200			180,378

Year	(1)	(2)	(3)	(4)	(5)	(6)	(7)	(8)	(9)	Total
1882	70,000	77,344	88,172	230,000	11,270	500,000	95,500			264,942
1883	20,000	386,719	429,258	1,170,000	50,310	400,000	66,000			565,568
1884	10,500	572,344	635,302	3,000,000	111,000	363,125	47,206			804,008
1885	10,000	900,000	963,000	4,400,000	171,600	400,000	43,200			1,187,800
1886	26,241	993,867	983,928	3,208,000	147,568	400,000	44,400			1,202,137
1887	22,853	952,255	933,210	2,668,135	120,066	666,000	91,908			1,168,037
1888	24,289	789,396	742,032	3,259,904	143,436	579,100	97,289			1,007,046
1889	26,436	913,254	858,459	4,704,261	183,466	397,804	53,704			1,122,065
1890	353,133	2,791,626	2,931,207	4,228,803	190,296	665,754	103,858			3,578,494
1891	478,750	2,273,054	2,250,323	4,128,887	179,262	865,044	110,726			3,019,061
1892	138,688	754,114	656,079	8,012,729	320,509	638,875	74,109			1,189,385
1893	188,854	1,221,155	952,501	8,000,000	296,000	600,000	64,800			1,502,155
1894	178,138	995,153	626,946	4,422,000	145,926	600,000	57,000			1,008,010
1895	172,697	1,515,693	985,200	5,747,003	183,904	600,000	64,200			1,406,001
1896	141,046	2,371,912	1,612,900	6,599,143	197,974	217,310	23,469			1,975,389
1897	552,840	2,776,394	1,665,836	7,784,212	280,232	2,185,084	262,210	20,000	$ 880	2,761,118
1898	852,555	2,346,194	837,995	2,799,936	106,398	1,035,565	128,410			1,925,358
1899	1,694,940	2,024,194	1,407,716	7,556,386	340,037	305,177	52,185			3,494,878
1900	1,437,909	1,985,267	1,230,866	9,478,657	417,061	352,368	58,493			3,145,209
1901	1,546,323	1,633,725	980,235	7,904,724	339,903	652,937	64,040			2,975,501
1902	2,420,726	789,855	418,623	4,262,063	174,745	526,541	64,238			3,078,332
1903	2,171,508	417,343	225,365	3,350,569	140,724	380,409	52,116			2,589,713
1904	2,174,361	294,028	170,536	2,044,525	87,915	431,048	55,174	5,016	256	2,488,242
1905	2,333,282	758,107	462,445	5,398,264	251,368	524,199	81,775	48,267	2,848	3,131,718
1906	992,179	916,256	623,054	5,721,599	326,131	662,111	127,784	10,377	633	2,069,784
1907	2,415,049	352,519	232,663	3,606,699	191,155	908,675	181,735	30,407	1,794	3,022,396
1908	2,028,698	415,070	219,987	3,033,352	127,401	1,019,574	134,584			2,510,670
1909	3,044,880	345,815	179,824	2,813,932	120,999	984,269	127,955	19,148	1,034	3,474,637
1910	2,195,847	414,250	223,695	4,004,728	176,208	620,326	78,770			2,674,520
1911	1,952,958	512,800	271,784	3,949,822	177,742	564,273	70,534			2,473,018
1912	1,049,590	545,177	335,284	2,989,044	134,507	400,552	66,091	140,667	9,706	1,595,178
1913	959,377	537,634	324,731	2,180,591	95,946	500,329	77,551	200,429	11,224	1,468,829
1914	1,211,993	594,289	328,642	2,199,564	82,663	854,038	113,587	44,608	2,275	1,739,160
1915	1,118,016	576,621	292,345	1,990,681	93,562	863,851	151,174	7,282	903	1,656,002
1916	491,175	803,461	528,677	2,339,029	161,393	444,081	109,244	69,015	9,248	1,299,737
1917	92,831	868,097	715,312	2,031,721	174,728	179,553	49,018	532,794	54,345	1,086,234
Totals	$34,675,607	37,043,150	$27,588,110	151,757,744	$6,482,595	22,387,879	$3,207,240	1,128,010	$95,146	$72,048,698
*1918	60,000	309,650	493,199	2,565,197	191,937	155,405	38,183			783,319

*Estimate.

Compiled from figures by C. W. Henderson, Statistician, U. S. Geological Survey, and from other sources.

TOTAL MINERAL PRODUCTION BY YEARS FOR PARK COUNTY

Yrs	GOLD Value	SILVER Fine Ounces	SILVER Value	LEAD Pounds	LEAD Value	COPPER Pounds	COPPER Value	ZINC Pounds	ZINC Value	Total
1867	$2,490,000		$ 20, 00	5, 00	- 00					$ 2,490,000
68	50,000									50,000
69	40,000									40,000
70	80,000									80,000
71	40,000	15, 04								60,300
82	50,000	142,209	188,000	50,000	3,200	69, 93	$ 47, 88			241,200
83	80,000	307,633	399,000	111,400	6,684	63, 91	44, 76			533,142
84	116,497	333,764	426,550			72, 50	16, 28			587,793
85	104,302	412,022	510, 07	25,000	1,450	68, 33	14, 50			633,039
86	60,000	386,719	448,594	50,000	3,050					525,994
87	60,000	309,375	371,250	150,000	8,250	170,000	32,300			471,800
88	60,000	309,375	355,781	150,000	5,400	175,000	29,050			450,231
89	60,000	324,844	363,825	300,000	12,300	100,000	37,200			473,325
90	50,000	293,906	337,992	300,000	15,000	100,000	42,800			445,792
91	50,000	270,703	305,894	312,000	14,976	100,000	18,200			389,070
82	100,000	193,359	220,429	312,000	15, 88	100,000	19,100			354,817
83	200,000	135,352	150,241	312,000	13,416					363,657
84	60,000	193,359	214,628	398,066	14, 28					289,356
85	60,000	71,310	76,302	398,066	15, 85					151,827
86	148,284	71,310	70,597	624,000	28, 04					247,585
87	648,462	07,513	105,363	708,713	31,892					785,717
88	33,945	50, 07	423,430	7,641,720	336,236					793,611
89	124,745	24, 73	211,258	4,640,682	180,487	855	115			517,105
90	37,281	56, 95	164,824	1,886,504	84,893					286,998
91	50,333	85, 80	183,345	19,656	85					234,526
82	39,687	43,792	38,099	25,698	1,028	10,000	1,088			78,814
83	109,845	62,350	48,633	30,060	1,110	10,000	90			160,668
84	97,358	43,817	27,605	30,000	90					126,903
85	131,761	46,658	30,328	98,791	3,161	2,938	314			165,564
86	137,108	117,095	79,625	297,714	8,931	28,593	3,088			228,753
87	153,619	199,945	119,965	4,517,614	162,634	58,002	6,962			443,182
88	159,490	198,711	117,239	1,953,001	74,214	20,957	2,599			353,542
89	153,041	72,137	43,282	540,849	24,338	7,903	1,351			222,012
90	116,558	43,138	26,746	682,107	30,113	15,000	2,490			175,807
01	96,322	69,175	41,505	421,955	18,144	9,657	1,613			157,584

TOTAL MINERAL PRODUCTION BY YEARS FOR PITKIN COUNTY

Years	GOLD Value	SILVER Fine Ounces	SILVER Value	LEAD Pounds	LEAD Value	COPPER Pounds	COPPER Value	ZINC Pounds	ZINC Value	Total
1880	$	10,000	$ 11,500	60,000	$ 3,000					$ 14,500
1881	$100,000	23,203	26,219	200,000	9,600					135,819
1882	90,000	23,203	26,451	200,000	9,800					126,251
1883	2,000	42,539	47,218	450,000	19,350					68,568
1884	1,000	464,062	515,109	1,700,000	44,400					560,509
1885	1,000	1,000,000	1,070,000	5,950,000	232,050					1,303,050
1886	17,125	399,094	395,103	800,000	36,800					449,028
1887	9,336	612,368	600,121	361,388	16,262					625,719
1888	12,716	4,333,787	4,073,760	14,349,792	631,391					4,717,867
1889		5,982,238	5,623,304	15,100,807	588,931					6,212,235
1890		4,944,898	5,192,143	19,703,605	886,662					6,078,805
1891	13,507	6,978,263	6,209,470	16,396,580	705,053					7,628,030
1892		8,138,549	7,080,638	20,998,701	839,948					7,920,486
1893		5,039,799	3,931,043	15,000,000	555,000					4,486,043
1894	5,312	5,996,851	3,778,016	15,750,000	519,750					4,303,078
1902	142,458	49,968	26,483	261,046	10,703	8,113	990			180,634
1903	136,277	52,128	28,149	802,489	33,105	5,895	808			198,939
1904	194,980	50,013	29,008	757,703	32,581	5,420	758			257,327
1905	320,867	49,202	30,013	543,303	25,535	12,199	1,903			378,318
1906	395,050	144,815	98,474	966,193	55,073	14,399	2,779			551,376
1907	513,216	111,215	93,402	1,062,732	56,325					642,943
1908	430,808	12,047	6,385	495,985	20,831	37,106	4,898	728,000	34,216	497,138
1909	551,921	102,375	53,235	2,237,093	96,195	61,023	7,933	366,574	19,795	729,079
1910	265,547	117,037	63,200	2,041,204	89,813	88,748	11,272	659,796	35,628	465,460
1911	58,882	69,072	36,600	923,089	41,538	24,216	3,027	407,772	23,243	163,249
1912	67,981	31,234	19,209	167,756	7,549	10,321	1,703	132,275	9,127	105,569
1913	50,441	94,293	56,953	506,046	22,266	29,161	4,520	98,623	5,523	139,303
1914	67,485	20,215	11,179	168,154	6,558	8,023	1,067	57,940	2,955	89,244
1915	159,339	9,227	4,678	190,830	8,969	12,303	2,153	472,992	58,651	233,790
1916	234,299	13,231	8,706	330,609	22,812	22,598	5,559	47,560	6,373	277,749
1917	117,358	14,705	12,117	278,709	23,969	12,824	3,501			156,945
Totals	$9,755,098	6,732,817	$6,679,041	38,725,477	$1,672,110	1,975,121	$375,014	2,971,532	$195,512	$18,676,775
1918*	56,000	18,999	18,284	224,467	16,714	11,115	2,731			93,729

*Estimate.

Compiled from figures by C. W. Henderson, Statistician, U. S. Geological Survey, and from other sources.

TOTAL MINERAL PRODUCTION BY YEARS FOR PITKIN COUNTY—Continued

Years	GOLD Value	SILVER Fine Ounces	SILVER Value	LEAD Pounds	LEAD Value	COPPER Pounds	COPPER Value	ZINC Pounds	ZINC Value	Total
1895	$ 1,387	5,131,792	$ 3,335,665	11,163,685	$ 357,238	612	$ 66	21,000	$ 756	$ 3,695,112
1896	1,523	4,922,360	3,347,205	16,272,411	483,172	52,991	5,723	......	......	3,842,623
1897	164,430	4,599,946	2,759,968	4,466,478	160,433	8,360	1,003	......	......	3,085,834
1898	71,001	3,977,270	2,346,589	15,903,682	604,340	4,553	565	......	......	3,022,495
1899	52,233	4,158,708	2,495,275	25,458,380	1,145,627	19,351	3,309	......	......	3,696,395
1900	13,456	4,119,116	2,553,852	27,452,260	1,207,899	6,082	1,010	20,000	880	3,777,097
1901	4,692	3,532,863	2,119,718	32,749,511	1,408,229	50,786	8,481	......	......	3,541,120
1902	4,899	3,063,450	1,623,629	24,973,816	1,023,926	10,654	1,300	......	......	2,653,754
1903	4,754	2,569,862	1,387,725	33,269,852	1,397,334	11,683	1,601	......	......	2,791,414
1904	2,336	2,129,618	1,235,178	18,882,901	811,965	9,862	1,262	593,661	30,277	2,081,018
1905	248	2,469,520	1,506,407	22,386,142	1,052,149	127,094	19,827	3,854,339	227,406	2,806,037
1906	1,172	2,131,374	1,449,334	17,951,674	1,023,245	285,346	55,072	3,276,711	199,899	2,728,702
1907	579	1,719,446	1,134,834	12,235,230	648,467	234,493	46,899	4,688,693	276,633	2,017,412
1908	538	1,041,700	552,101	7,568,060	317,859	22,474	2,967	727,362	33,951	907,416
1909	745	700,038	364,020	13,143,210	565,158	26,092	3,392	34,741	1,876	935,191
1910	646	477,813	258,019	13,408,250	589,963	24,843	3,155	......	......	851,783
1911	542	450,772	238,909	11,084,334	498,795	7,408	926	......	......	739,172
1912	165	528,504	325,030	8,405,333	378,240	22,952	3,787	484,507	33,431	740,653
1913	29	562,308	339,634	17,528,386	771,249	48,852	7,572	460,161	25,769	1,144,253
1914	123	371,886	206,206	23,233,230	906,096	67,737	9,009	145,141	7,417	1,129,151
1915	29	448,915	227,600	19,265,213	905,465	19,983	3,497	214,952	26,654	1,163,245
1916	......	577,863	380,234	17,519,275	1,208,830	28,931	7,117	162,574	21,785	1,617,916
1917	105	662,045	545,525	14,352,523	1,234,317	27,403	7,481	571,794	58,323	1,845,751
Totals ..	$577,928	94,338,023	$70,012,602	535,184,709	$23,802,993	1,118,546	$195,021	15,250,926	$945,037	$95,533,581
*1918	......	571,904	553,390	11,666,621	872,597	10,215	2,505	150,408	11,890	1,440,382

*Estimate.

Compiled from figures by C. W. Henderson, Statistician, U. S. Geological Survey and from other sources.

TOTAL MINERAL PRODUCTION BY YEARS FOR RIO GRANDE COUNTY

Years	GOLD Value	SILVER Fine Ounces	SILVER Value	LEAD Pounds	LEAD Value	COPPER Pounds	COPPER Value	ZINC Pounds	ZINC Value	Total
1870-1872	No production.									
1873	$ 2,000	......	......	......	......	......	......	......	......	2,000
1874	5,000	......	......	......	......	......	......	......	......	5,000

Year	(1)	(4)	(5)	(6)	(7)	(8)	(9)	(10)
1875	281,634					$ 9,590	7,734	272,044
1876	130,119					8,971	7,734	121,148
1877	204,618					9,281	7,734	195,337
1878	111,760					8,894	7,734	102,866
1879	37,162					8,662	7,734	28,500
1880	6,000					8,739	7,734	6,000
1881	298,739					17,635	15,469	290,000
1882	227,635					8,585	7,734	210,000
1883	188,585					12,019	10,828	180,000
1884	142,019							130,000
1885	140,486					10,486	9,800	130,000
1886	157,375					8,729	8.8	149,266
1887	130,212					7,832	7,992	122,380
1888	19,008					2,748	2,923	16,260
1889	39,292					3,532	3,757	35,760
1890	27,067					1,351	1,287	25,716
1891	46,266					7,674	7,752	38,592
1892	25,385					10,898	12,526	14,487
1893	621					621	786	
1894	17,610					794	1,260	16,816
1895	17,978	$ 148	1,369	$ 14	451	2,183	3,359	15,785
1896	2,952	75	87	432	12,006	920	1,353	1,870
1897	28,000	1,214	9,794	91	2,393	4,901	8,168	22,592
1898	5,950	57	36	74	1,635	925	1,568	3,720
1899	20,964					1,631	2,718	19,202
1900	112,117	1,427	8,599	1,155	26,260	$ 905	3,075	107,629
1901	48,068	10,956	65,603	29	677	4,156	6,926	32,927
1902	16,104	154	1,260	7	166	1,681	3,171	14,262
1903	15,478	698	5,098			1,841	3,410	12,939
1904	5,416	83	650			1,323	2,281	4,010
1905	4,714	19	123			644	1,055	4,051
1906	8,683					103	152	8,580
1907				11				
1908	764				250	33	61	764
1909								
1910	1,361	4,896	87	11	313	551	896	1,306
1911	11,010	88	29,673	14		66	109	5,549
1912	397		568			9	16	243
1913	483							474
1914								

TOTAL MINERAL PRODUCTION BY YEARS FOR RIO GRANDE COUNTY—Continued

Years	GOLD Value	SILVER Fine Ounces	SILVER Value	LEAD Pounds	LEAD Value	COPPER Pounds	COPPER Value	ZINC Pounds	ZINC Value	Total
1915	$ 14,968	325	$ 165							$ 15,133
1916										
1917	113	3,170	2,612	2,651	228	330	90			3,043
Totals	$2,363,166	179,158	$172,691	46,802	$2,055	124,117	$19,916			
1918										$2,557,828

Compiled from figures by C. W. Henderson, Statistician, U. S. Geological Survey, and from other sources.

TOTAL MINERAL PRODUCTION BY YEARS FOR ROUTT AND MOFFAT COUNTIES

Years	GOLD Value	SILVER Fine Ounces	SILVER Value	LEAD Pounds	LEAD Value	COPPER Pounds	COPPER Value	ZINC Pounds	ZINC Value	Total
1866-1872	No production.									
1873	$ 26,000									$ 26,000
1874	No production.									
1875	No production.									
1876-1880	3,500									3,500
1881	20,000									20,000
1882	15,000									15,000
1883	40,000									40,000
1884	13,000									13,000
1885	23,000									23,000
1886	16,840	387	$ 383							17,223
1887	6,714	214	210							6,924
1888	No production.									
1889	8,870	189	178							9,048
1890	633	176	185							35.8
1891	13,561									13,561
1892	560									560
1893	25.6					958	$ 115			6,216
1894	8,944	97	61			600	74			6.5
1895	5,930	86	56							5,986
1896	4,859	2,214	1,506	22,111	$ 663					7,028
1897	9,777	7,805	4,683	88,736	3,194					17,769
1898	12,753	2,173	1,282	15,477	588					14,697
1899	11,555	1,271	763	3,405	153					12,471
1900	3,287	477	296			5,765	957			4,540

TOTAL MINERAL PRODUCTION BY YEARS FOR SAGUACHE COUNTY

Years	GOLD Value	SILVER Fine Ounces	SILVER Value	LEAD Pounds	LEAD Value	COPPER Pounds	COPPER Value	ZINC Pounds	ZINC Value	Total
1880		7,734	$ 8,894							$ 8,894
881	$ 10,000	30,938	34,960							34,960
882	5,000	77,344	88,172							9,672
1883	1,000	77,344	85,852							90,852
1884		77,344	85,852							86,852
8851	1,000	55,920	59,834							60,834
1886	3,936	55,920	55,361							59,297
1887	756	7,196	7,052	12,582	$ 566					8,374
1888	4,220	36,101	33,935	180,272	7,932					46,087
1889				200,000	7,800					7,8 0
89	1,745	11,988	12,587	176,193	7,929	4,290	$ 669			22,930
891	1,422	21,285	21,072	260,571	11,205	68,047	8,710			42,4 0
1892				250,000	10,000					10,000
1893				250,000	9,250					9,250
1894	17,515	608,224	383,181	250,000	8,250					408,946

*Estimate.
Compiled from figures by C. W. Henderson, Statistician, U. S. Geological Survey, and from other sources.

Years	GOLD Value	SILVER Fine Ounces	SILVER Value	LEAD Pounds	LEAD Value	COPPER Pounds	COPPER Value	ZINC Pounds	ZINC Value	Total
9 0	4,444	239	143							4,765
9 2	51	136	72							15,223
9 3	20,835	117	63							20,898
1904	24,225	181	105							24,330
1905	55	30	18	2,193	94	500	84			6,923
9 6	6,95	42	29							6,980
9 7	5, 0	429	283							5,292
9 8	5,2 0	1,242	658							
9 9	3,361	3,440	1,792							
0 0	00	48	26							6,715
9 1	6,115	47	25							6,140
9 2	50	150	92							9,301
9 3	3,840	1,9 02	1,185	1,023	45	25,085	4,139			0 5
9 1	4,697	16	9			161	25			4,706
9 5	2,984	2	6							2,990
9 0	1, 2	278	183			41,175	10,129			11,454
9 1	3,415	1,341	6			4,326	1,181			5,701
Totals	$384,539	24,805	$15,397	132,945	$4,737	78,570	$16,704			$421,377
1918*		2,800	2,709							3,709

TOTAL MINERAL PRODUCTION BY YEARS FOR SAGUACHE COUNTY—Continued

Years	GOLD Value	SILVER Fine Ounces	SILVER Value	LEAD Pounds	LEAD Value	COPPER Pounds	COPPER Value	ZINC Pounds	ZINC Value	Total
1895	$534	3,939	$2,560	249,166	$7,973		$			11,067
1896	331	2,447	1,664	65,465	1,964	241	26			3,985
1897	13,746	2,482	1,489	9,266	334	2,975	375			15,926
1898	19,678	2,618	1,545	132,462	5,034	21,711	2,692			28,949
1899	3,886	14,306	8,584	441,095	19,849	35,319	6,040			38,359
1900	7,979	15,793	9,792	316,061	13,907	16,129	2,677			34,355
1901	79,972	20,507	12,304	235,750	57	15,253	2,547			94,960
1902	5,023	10,486	5,558	454,995	18,655	13,669	1,668	267,100	$12,821	43,725
1903	2,956	22,424	12,090	376,711	15,822	67,410	9,235	44,600	2,408	42,530
1904	5,519	60,506	35,093	699,312	30,070	48,722	6,236	15,585	795	77,713
1905	699	4,401	2,685	203,797	9,578	1,135	177			351
1906	7,628	737	501	49,141	2,801			2,917	172	15,462
1907	649	6,194	4,088	22,528	1,194	1,260	252	74,302	4,532	6,183
1908	0	953	505	27,715	1,164	76	10			2,289
1909	1,196	2,260	1,175	83,463	3,589	3,769	490			6,450
1910	1,025	4,841	2,614	161,068	7,087	5,362	681	46,561	2,654	11,407
1911	512	4,664	2,472	74,556	3,355	4,984	625			9,636
1912	3,805	19,309	11,875	504,845	22,718	29,479	4,864	534,928	36,910	80,172
1913	4,243	8,694	5,251	336,886	14,823	13,277	2,058	32,964	1,846	28,221
1914	16,513	18,293	10,116	534,872	20,860	35,783	4,759	8,941	456	52,704
1915	5,273	11,266	5,712	174,447	8,199	23,360	4,088	44,250	5,487	28,759
1916	8,024	48,959	32,215	255,449	17,626	92,581	22,775			80,640
1917	10,261	72,898	60,068	309,965	26,657	144,648	39,489			136,475
Totals	$246,656	1,426,315	$1,106,727	7,298,639	$326,328	649,480	$121,123	1,072,138	$68,081	$1,868,915
*1918	3,000	86,410	83,602	119,090	8,831	113,873	27,954			123,387

*Estimate.
Compiled from figures by C. W. Henderson, Statistician, U. S. Geological Survey, and from other sources.

TOTAL MINERAL PRODUCTION BY YEARS FOR SAN JUAN COUNTY.

Years	GOLD Value	SILVER Fine Ounces	SILVER Value	LEAD Pounds	LEAD Value	COPPER Pounds	COPPER Value	ZINC Pounds	ZINC Value	Total
1873	$13,000		$		$		$		$	13,000
1874	9,540	3,166	4,046							13,586
1875	10,000	68,547	84,998	120,000	6,960					101,958
1876	5,000	48,465	56,219	249,348	15,210					76,429
1877	5,000	34,010	40,812	400,000	22,000	8,664	1,646			69,458

Year	1	2	3	4	5	6	7	8	9	10
1878	54,654			6,000	36,145	14,400	400,000	28,254	24,569	6,000
1879	79,751			18,600	100,000	20,500	500,000	34,651	30,938	6,000
1880	62,212			18,400	100,000	21,500	430,000	13,342	11,602	6,000
1881	51,770			18,200	100,000	6,720	140,000	21,850	19,336	5,000
1882	97,683			19,100	100,000	15,680	320,000	52,903	46,406	10,000
1883	400,871			16,500	100,000	48,891	1,137,000	300,480	270,703	35,000
1884	719,909			39,000	300,000	125,800	3,400,000	515,109	464,062	40,000
1885	1,006,500			10,800	100,000	206,700	5,300,000	749,000	700,000	40,000
1886	1,063,037			11,100	100,000	197,800	4,300,000	711,338	718,523	142,799
1887	648,176			41,400	300,000	91,806	2,040,145	393,725	401,760	121,245
1888	545,411			40,320	240,000	104,824	2,382,358	209,939	223,339	190,328
1889	1,050,707			18,227	135,018	159,779	4,096,887	477,828	508,328	394,873
1890	703,548			22,987	147,354	155,797	3,462,153	337,407	321,340	187,357
1891	1,278,973			30,140	235,467	294,874	6,857,544	761,850	769,545	192,109
1892	766,943			15,865	136,768	256,267	6,406,665	345,903	397,589	148,908
1893	933,436			121,589	1,125,826	296,000	8,000,000	255,179	357,153	260,668
1894	799,456			106,231	1,118,222	132,000	8,000,000	221,202	351,114	340,023
1895	2,560,129			229,162	2,057,588	259,162	8,098,800	1,231,394	1,894,463	841,411
1896	2,684,076			91,270	845,094	169,038	5,634,586	1,515,061	2,228,031	908,707
1897	1,816,465			172,224	1,435,303	288,771	8,021,414	661,144	1,101,987	694,326
1898	2,587,586			279,300	2,252,421	557,080	14,659,999	618,614	1,048,499	1,132,592
1899	2,636,712			204,800	1,197,661	720,525	16,011,617	715,114	1,191,857	996,273
1900	2,280,471			321,867	1,972,087	773,484	17,579,177	422,416	681,317	757,204
1901	2,556,439			457,587	2,740,042	665,347	15,473,187	470,531	784,218	962,974
1902	2,651,614			367,499	3,012,283	315,695	7,699,883	444,194	838,102	1,524,226
1903	2,827,588			402,645	2,939,018	292,702	6,969,093	421,933	1,042,044	1,710,608
1904	2,860,421	16,180	317,254	433,792	3,467,124	399,412	9,288,643	604,386	1,042,691	1,396,651
1905	2,251,535	9,667	163,845	354,791	2,274,109	378,121	8,045,126	458,015	750,204	1,050,971
1906	1,969,695	43,810	718,192	299,085	1,549,663	257,373	4,515,317	469,252	690,076	900,175
1907	2,999,623	104,593	1,772,764	490,056	2,450,280	661,626	12,483,507	775,616	1,175,176	967,732
1908	2,184,801	476	10,131	301,321	2,282,738	352,908	8,402,569	532,272	1,004,287	997,824
1909	1,744,003	42,472	786,518	214,915	1,653,192	390,658	9,085,088	412,691	793,637	683,267
1910	1,960,898	204,188	3,781,259	153,479	1,208,496	470,289	10,688,386	422,415	732,250	710,527
1911	1,006,707	126,788	2,224,351	58,864	470,912	312,022	6,933,822	172,570	325,604	336,463
1912	1,719,894	171,023	2,478,594	175,443	1,063,291	410,145	9,114,334	439,709	714,974	523,574
1913	1,890,349	93,240	1,664,999	189,335	1,221,516	418,395	9,508,979	531,767	880,409	657,612
1914	1,143,653	49,530	971,177	109,749	825,180	202,761	5,199,000	273,136	493,917	508,477

TOTAL MINERAL PRODUCTION BY YEARS FOR SAN JUAN COUNTY—Continued

Years	GOLD Value	SILVER Fine Ounces	SILVER Value	LEAD Pounds	LEAD Value	COPPER Pounds	COPPER Value	ZINC Pounds	ZINC Value	Total
1915	$ 583,681	430,637	$ 218,333	6,791,596	319,205	1,054,463	$ 184,521	2,259,226	$ 280,114	$ 1,585,894
1916	438,628	502,342	330,541	7,285,304	502,686	1,615,167	397,331	4,014,403	537,930	2,207,116
1917	318,006	658,261	542,407	10,515,535	904,336	1,665,923	454,797	3,270,500	333,591	2,553,137
Totals	$21,778,759	26,534,675	$18,299,546	271,947,107	$12,215,249	45,736,915	$6,909,418	24,433,213	$2,013,632	$61,216,604
1918*	250,000	430,270	418,928	8,943,139	668,896	1,087,743	267,013	3,213,100	254,000	1,853,837

*Estimate.

Compiled from figures by C. W. Henderson, Statistician, U. S. Geological Survey, and from other sources.

TOTAL MINERAL PRODUCTION BY YEARS FOR SAN MIGUEL COUNTY

Years	GOLD Value	SILVER Fine Ounces	SILVER Value	LEAD Pounds	LEAD Value	COPPER Pounds	COPPER Value	ZINC Pounds	ZINC Value	Total
1875		3,867	$ 4,795							$ 4,795
1876		3,867	4,486							4,486
1877		3,867	4,640							4,640
1878	$ 5,000	3,867	4,447	50,000	$ .80					11,247
1879	7,000	7,734	8,662	350,000	14,350					30,012
1880	5,000	7,734	8,894	482,500	24,125					38,019
1881	15,000	19,336	21,850	200,000	9,600					46,450
1882	10,000	38,672	44,086	200,000	9,800					63,886
1883	125,000	193,359	214,628	782,000	33,626					373,254
1884	100,000	309,375	373,406	300,000	11,100					454,506
1885	100,000	400,000	428,000	300,000	11,700					539,700
1886	217,570	430,805	426,497	300,000	13,800					657,867
1887	169,696	492,725	482,871	537,144	24,171					676,738
1888	424,706	663,354	623,553	636,514	28,007					1,076,266
1889	432,588	726,456	682,869	1,166,346	45,488					1,160,945
1890	755,380	907,148	952,505	414,522	18,653					1,726,538
1891	646,993	1,410,903	1,397,525	139,344	5,992					2,050,510
1892	694,177	1,501,898	1,306,651	815,842	32,634	100,000	$ 11,600			2,045,062
1893	682,680	932,568	727,403	7,0	25,900	200,000	21,600			1,467,583
1894	794,218	570,023	359,115	858,830	28,341	173,191	16,453			1,198,127
1895	1,426,159	602,039	391,325	756,809	24,218	147,727	15,807			1,857,509
1896	1,377,829	1,109,875	754,715	2,284,191	68,526	21,698	2,343			2,203,413
1897	1,548,144	869,079	521,447	4,143,767	149,176	354,781	42,574			2,171,341
1898	1,572,667	2,129,082	1,256,158	6,699,712	254,589	360,831	44,743			3,128,167
1899	1,376,705	1,208,395	725,037	3,918,883	176,350	160,239	27,401			2,305,493

TOTAL MINERAL PRODUCTION BY YEARS FOR SUMMIT COUNTY

Years	GOLD Value	SILVER Fine Ounces	SILVER Value	LEAD Pounds	LEAD Value	COPPER Pounds	COPPER Value	ZINC Pounds	ZINC Value	Total
1900	1,827,352	1,136,692	704,749	3,353,425	47,551	311,045	51,633			2,731,285
1901	2,049,472	916,245	549,747	3,309,517	142,309	308,322	5,940			2,793,018
1902	2,007,656	1,056,640	560,019	4,296,849	76,710	454,784	55,484			2,799,330
1903	1,176,805	737,028	397,995	3,704,201	155,576	466,264	63,878			1,794,254
1904	1,531,068	667,710	387,273	5,704,708	246,302	239,520	30,659		1,016	2,194,301
1905	1,711,853	1,275,079	777,798	6,970,152	327,597	272,513	42,512		17,214	2,860,776
1906	2,447,709	1,437,389	1,207,340	7,158,189	408,017	319,692	61,701			4,054,823
1907	2,467,516	1,438,299	949,277	6,499,957	344,986	381,437	76,287			3,837,578
1908	2,317,651	1,543,187	817,889	7,135,863		562,888	74,301	952,872	44,785	3,554,332
1909	2,285,051	1,344,152	698,959	4,941,370	212,479	501,285	65,167	804,296	43,432	3,305,088
1910	2,497,793	1,144,050	617,787	7,791,841	342,841	544,189	69,112			3,643,008
1911	2,447,841	1,000,834	530,442	6,456,333	290,535	971,064	121,383	2,193,981	118,475	3,583,208
1912	2,294,304	1,153,709	709,531	7,429,622	334,333	845,497	139,507	3,386,008	193,007	3,785,726
1913	2,129,371	1,051,096	634,862	6,967,136	306,554	736,374	114,138	2,943,783	203,121	3,319,647
1914	2,114,196	1,280,461	708,095	4,039,769	157,551	324,105	43,106	2,405,750	134,722	3,023,568
1915	2,069,362	1,096,641	555,097	5,240,277	246,293	562,554	98,447	1,040,121	128,975	3,099,074
1916	2,072,393	812,041	534,323	6,125,551	422,732	581,427	81,000	1,098,485	147,197	3,319,676
1917	2,009,961	779,364	642,196	6,205,326	533,658	920,425	251,276	1,810,245	184,645	3,621,736
Totals	$49,956,607	34,651,778	$23,609,818	129,367,490	$6,105,649	10,821,858	$1,735,633	16,652,835	$1,199,375	$82,607,082
*1918	2,170,000	1,168,250	1,131,783	5,906,906	441,803	1,195,915	269,021	493,450	39,000	4,051,607
1860-1867	$5,150,000									$5,150,000
1868	150,000									150,000
1869	200,000	7,547	$10,000	50,000	$3,000					213,000
1870	500,000	7,907	10,500	50,000	3,000					513,600
1871	70,000									70,000
1872	120,000	3,782	5,000	100,000	6,000					131,400
1873	75,000	3,855	5,000	100,000	6,000					86,000
1874	70,000	23,784	30,396	423,950	25,437					125,833
1875	72,012	7,734	9,590	141,000	8,178					89,780
1876	160,000	154,688	179,438	100,000	6,100					335,638

*Estimate.
Compiled from figures by C. W. Henderson, Statistician, U. S. Geological Survey, and from other sources.

TOTAL MINERAL PRODUCTION BY YEARS FOR SUMMIT COUNTY—Continued

Years	GOLD Value	SILVER Fine Ounces	SILVER Value	LEAD Pounds	LEAD Value	COPPER Pounds	COPPER Value	ZINC Pounds	ZINC Value	Total
1877	150,000	30,938	37,126	100,000	5,550					192,626
1878	165,774	119,883	137,865	100,000	3,600					307,239
1879	75,000	154,688	173,251	100,000	4,100					252,351
1880	49,000	317,109	364,675	500,000	25,000					438,675
1881	31,000	1,560,344	1,763,189	16,773,000	805,104					2,599,293
1882	55,000	674,757	769,223	5,773,000	282,877					1,107,100
1883	15,000	270,703	300,480	2,773,000	119,239					434,719
1884	20,000	232,031	257,554		36,454					314,008
1885	200,000	234,351	250,756	985,250	38,425			25,000	1,075	490,256
1886	164,222	422,298	418,075	1,546,000	71,116			25,000	1,100	654,513
1887	240,520	220,120	215,718	1,745,132	78,936			25,000	1,150	536,324
1888	282,209	394,058	370,415	2,126,887	93,583			75,000	3,750	749,882
1889	222,724	519,874	488,651	3,055,981	119,183	2,066	278	75,000	3,750	834,586
1890	222,724	519,842	488,651	3,055,981	119,183			75,000	4,125	1,126,040
1891	89,132	523,658	518,421	10,591,152	455,420			75,000	3,750	1,066,723
1892	126,046	563,417	490,173	6,371,637	254,865	166,799	19,349	212,500	9,775	900,208
1893	116,168	421,566	328,821	5,277,000	232,249			415,000	16,600	693,838
1894	224,791	432,794	272,660	5,500,000	181,500	1,058	113	200,000	7,000	685,951
1895	235,591	288,242	187,357	5,477,117	175,268			65,000	2,340	600,669
1896	210,202	441,448	221,332	3,950,010	118,501	54,081	5,841	100,000	3,900	638,629
1897	273,650	514,107	308,464	1,748,761	62,955	133,482		82,489	3,382	664,469
1898	343,825	415,687	245,255	4,889,204	185,790	9,825		227,156	10,449	786,537
1899	260,566	264,872	158,923	4,032,431	181,459	65,531	11,205	125,416	7,274	619,427
1900	338,182	403,330	250,065	4,610,710	246,871	53,030		491,055	21,606	865,527
1901	338,719	368,887	221,332	4,342,437		17,062	2,849	1,000,000	41,000	790,625
1902	242,583	274,571	145,523	3,092,387	126,788	93,609	11,420	1,329,180	63,801	590,115
1903	222,265	220,543	119,093	1,523,703	63,996	41,447	5,7..	550,800	29,743	440,775
1904	208,126	8,5..	104,721	2,178,182	93,662	7,510	961	1,884,584	96,114	503,584
1905	157,476	209,356	127,707	2,181,660	102,538	44,033	.8..	3,320,237	195,894	590,484
1906	139,773	107,752	73,271	1,301,912	74,209	27,120	5,234	3,363,740	205,188	497,675
1907	106,590	127,847	84,379	1,915,133	101,502	21,865	4,373	2,970,991	175,288	472,132
1908	186,941	6,2..	34,993	1,719,190	72,206	28,523	3,95	1,232,148	57,911	355,816
1909	452,766	99,763	51,877	3,559,278	153,049	3,839		5,798,167	313,101	971,292
1910	368,761	152,250	82,2..	5,015,409	220,678	21,740	2,761	5,542,685	299,305	973,725
1911	284,241	182,957	96,967	6,024,867	271,119	22,888	2,861	7,675,175	437,485	1,092,673

Year	GOLD Value	SILVER Fine Ounces	SILVER Value	LEAD Pounds	LEAD Value	COPPER Pounds	COPPER Value	ZINC Pounds	ZINC Value	Total
1912	426,015	164,665	101,269	4,402,422	198,109	16,412	2,708	9,342,725	644,648	1,372,749
1913	462,228	167,490	101,164	3,944,268	173,548	18,170	2,816	6,931,074	388,140	1,127,896
1914	668,610	67,009	37,056	1,565,231	61,044	7,339	976	5,111,941	260,709	1,028,395
1915	680,144	64,223	32,561	1,916,298	90,066	8,646	1,513	8,597,411	1,066,079	1,870,368
1916	673,891	120,207	79,096	1,688,637	116,516	14,581	3,587	13,940,948	1,868,087	2,741,177
1917	603,437	175,699	144,776	915,535	78,736	25,033	6,834	19,868,814	2,026,619	2,860,402
Totals	$16,898,015	12,895,696	$11,037,402	147,047,816	$6,370,510	905,689	$128,529	100,754,237	$8,270,063	$42,704,519
*1918	475,000	114,612	112,334	720,494	53,889	14,871	3,654	16,834,500	1,330,000	1,974,877

*Estimate.

Compiled from figures by C. W. Henderson, Statistician, U. S. Geological Survey, and from other sources.

TOTAL MINERAL PRODUCTION BY YEARS FOR CRIPPLE CREEK DISTRICT

Years	GOLD Value	SILVER Fine Ounces	SILVER Value	LEAD Pounds	LEAD Value	COPPER Pounds	COPPER Value	ZINC Pounds	ZINC Value	Total
1891	$ 1,930		$							
1892	557,851									557,851
1893	2,021,088	5,680	4,430							2,025,518
1894	2,618,388	25,335	15,961							2,634,349
1895	6,166,144	68,428	44,478							6,210,622
1896	7,413,493	63,617	43,260							7,456,753
1897	10,131,855	59,879	35,927							10,167,782
1898	13,507,349	67,799	40,001							13,547,350
1899	16,058,564	82,299	49,379							16,107,943
1900	18,149,645	80,792	50,091							18,199,736
1901	17,234,294	89,560	53,736							17,288,030
1902	16,932,416	62,780	33,273							16,865,689
1903	11,840,272	41,605	22,467							11,862,739
1904	14,456,536	47,017	27,734							14,484,270
1905	15,641,754	56,951	34,740							15,676,494
1906	13,930,526	67,943	46,201							13,976,727
1907	10,370,284	51,630	34,076							10,404,360
1908	13,031,917	52,270	27,703							13,059,620
1909	11,466,227	63,204	32,866							11,499,093
1910	11,002,253	54,263	29,302							11,031,555
1911	10,552,653	57,783	40,625							10,593,278
1912	11,008,362	66,117	40,662							11,049,024
1913	10,905,003	71,349	43,095							10,948,098
1914	11,996,116	89,056	49,248							12,045,364
1915	13,683,494	87,767	44,498							13,727,992

TOTAL MINERAL PRODUCTION BY YEARS FOR CRIPPLE CREEK DISTRICT—Continued

Years	GOLD Value	SILVER Fine Ounces	SILVER Value	LEAD Pounds	LEAD Value	COPPER Pounds	COPPER Value	ZINC Pounds	ZINC Value	Total
1916	$12,119,550	79,804	$ 52,577							$ 12,172,061
1917	10,394,847	64,568	53,204							10,448,051
Totals	$293,202,811	1,558,296	$939,288							$294,142,099
1918*	8,300,000	52,068	50,379							8,350,379

*Estimate.
Compiled from figures by C. W. Henderson, Statistician, U. S. Geological Survey, and from other sources.

TOTAL MINERAL PRODUCTION BY YEARS FOR MESA COUNTY

Years	GOLD Value	SILVER Fine Ounces	SILVER Value	LEAD Pounds	LEAD Value	COPPER Pounds	COPPER Value	ZINC Pounds	ZINC Value	Total
1885	$ 431	3	$ 3							$ 434
1886	110									110
1887-1893										
1894	318	1	1							319
1895-897										
1898	165	20	12			4,650	$ 795			177
1899	124	4,120	2,472			2,150	357			3,391
190	124	311	193			7,795	1,302			674
902	2,046	115	93			15,000	1,830			3,441
	537	32	17							2,384
903	351	8	4							355
904	248	9	5							253
905										
90	473	15	10							483
907	76	3	2							78
1908-1910	No production.									
91 1	28									28
1912	9	257	158	20	$1	5,685	938			1,106
1913-1917	No production.									
1918	No production.									
Totals	$5,040	4,934	$2,970	20	$1	35,280	$5,222			$13,233

Compiled from figures by C. W. Henderson, Statistician, U. S. Geological Survey, and from other sources.

CPSIA information can be obtained
at www.ICGtesting.com
Printed in the USA
BVHW04*1010190918
527934BV00014B/766/P